The Physics of OSCILLATIONS and WAVES

Ingram Bloch

THE PHYSICS OF OSCILLATIONS and WAVES

With Applications in
ELECTRICITY
and MECHANICS

Plenum Press • New York and London

Library of Congress Cataloging-in-Publication Data

Bloch, Ingram.
The physics of oscillations and waves : with applications in electricity and mechanics / Ingram Bloch.
p. cm.
Includes bibliographical references and index.
ISBN 0-306-45721-0
1. Electricity--Mathematics. 2. Mechanics. 3. Oscillations. 4. Waves. I. Title.
QC522.B58 1997
530.4'16--DC21 97-15933
CIP

ISBN 0-306-45721-0

A Division of Plenum Publishing Corporation
233 Spring Street, New York, N. Y. 10013

http://www.plenum.com

10 9 8 7 6 5 4 3 2 1

Book Designed by Reuven Solomon

Printed in the United States of America

This book is dedicated to

MARY BLOCH

whose encouragement, advice,

and labor have been essential

to its completion

Preface

Except for digressions in Chapters 8 and 17, this book is a highly unified treatment of simple oscillations and waves. The phenomena treated are "simple" in that they are describable by linear equations, almost all occur in one dimension, and the dependent variables are scalars instead of vectors or something else (such as electromagnetic waves) with geometric complications. The book omits such complicated cases in order to deal thoroughly with properties shared by all linear oscillations and waves.

The first seven chapters are a sequential treatment of electrical and mechanical oscillating systems, starting with the simplest and proceeding to systems of coupled oscillators subjected to arbitrary driving forces. Then, after a brief discussion of nonlinear oscillations in Chapter 8, the concept of normal modes of motion is introduced and used to show the relationship between oscillations and waves. After Chapter 12, properties of waves are explored by whatever mathematical techniques are applicable. The book ends with a short discussion of three-dimensional problems (in Chapter 16), and a study of a few aspects of non-linear waves (in Chapter 17).

Besides trying to keep the geometry simple, I have made no attempt to discuss the use of an electronic computer on any of the equations in the book; the only cases in which such use would be justified are a few of those in Chapters 8 and 17. Also omitted are topics whose understanding would require a knowledge of either relativity theory or quantum theory.

One of the purposes of the book is to prepare students for the study of quantum mechanics. My experience has been that, in trying to learn new mathematics, students of quantum theory are likely to slight the new and difficult physics which is the main point of the course. Students have found it helpful to learn much of the mathematics in working on problems in classical physics—especially so when the classical problems have their own interest.

Students planning to use this book should have at least a little knowledge of linear differential equations and of complex numbers. Students with more extensive backgrounds can use some parts of the book (e.g., Chapters 1–5, 10) as review. For students who have not studied vector analysis, instructors may wish to reformulate in one dimension the three-dimensional topics in Chapter 16. Helped by such flexibility, physics majors who are juniors, seniors, and beginning graduate students should be adequately prepared to use the book.

It is important for physics students to work on problems. There are problems at the ends of chapters, and brief answers to a few of them in the back of the book. More problems can be found in some of the books cited in the notes at the ends of chapters.

I greatly appreciate the patient help that I have received from Sidney and Raymond Solomon.

Ingram Bloch

Contents

1

Undamped and Undriven Oscillators and LC Circuits

All but two chapters of this book deal with the behavior of physical systems that are describable by linear differential equations. The treatment proceeds from simpler to more complicated systems and uses successively more sophisticated techniques of analysis. In this chapter we deal with the simplest oscillating system.

Let us consider a mass m that moves without friction on the x-axis and is subject to a force toward the origin that is proportional to the distance of the mass from the origin. Thus the force is

$$F = -kx, \tag{1.1}$$

and Newton's second law implies that

$$m\, d^2x/dt^2 = m\ddot{x} = -kx, \tag{1.2}$$

or, if $k/m = \omega_0^2$,

$$\ddot{x} + \omega_0^2 x = 0. \tag{1.3}$$

Such a system, with the equation of motion 1.2 or 1.3, is known as an undamped, undriven harmonic oscillator.

A force like that in Equation 1.1 can be produced by any object that is described by Hooke's law—for instance, a coil spring on which the mass is hung—provided the origin is taken to be the position of the mass at equilibrium. More generally, any system in which a mass that moves in one dimension has a position of stable equilibrium can be approximately described by the foregoing equations, if the mass has a potential energy that is an analytic function of x in some neighborhood of the equilibrium point. Letting x_0 be the point of stable equilibrium, we require that the potential energy $U(x)$ have vanishing derivative (condition for equilibrium) and positive second derivative (condition for stability), at x_0:

$$\begin{aligned} U'(x_0) &= 0; \\ U''(x_0) &> 0. \end{aligned} \tag{1.4}$$

The function $U(x)$, assumed analytic in a neighborhood of x_0, can be represented within that neighborhood by a Taylor series about x_0:

$$U(x) = U(x_0) + (x - x_0)U'(x_0) + \tfrac{1}{2}(x - x_0)^2 U''(x_0) + \ldots \tag{1.5}$$

In studying small departures of the mass from its equilibrium position, one is dealing with small values of $|x - x_0|$ and can approximate the force acting on the mass by dropping the cubic and higher terms from the series 1.5. In that case, within the domain of analyticity of $U(x)$, the potential energy can be approximated as

$$U(x) \cong U(x_0) + \tfrac{1}{2}k(x - x_0)^2, \tag{1.6}$$

where we have omitted the vanishing first derivative and have replaced the second derivative of U by the positive constant k. Now, using the relationship between potential energy and force, we have

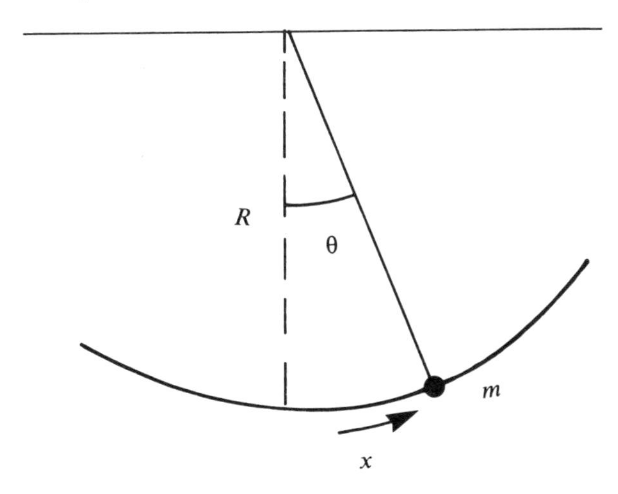

Figure 1.1. Simple pendulum swinging in a plane.

$$F = -dU/dx = -k(x - x_0). \tag{1.7}$$

In order to reproduce Equation 1.1, we have only to place the point of equilibrium at the origin, $x_0 = 0$, or, equivalently, to choose $x - x_0$ as a new variable, which choice does not affect the second time derivative in Equations 1.2 and 1.3.

The most familiar example of a system for which Equation 1.2 or Equation 1.3 is an approximate equation of motion, valid for small excursions from equilibrium, is the simple pendulum swinging in a plane (see Figure 1.1). Here x is taken to be measured along the arc on which the pendulum bob moves, with the origin at the position of equilibrium. Thus, if R is the length of the string and θ is the angle of the string from the vertical, $x = R\theta$, or $\theta = x/R$. The potential energy of the mass m is $U = mgh$, where g is the acceleration of gravity and h is the height of the mass above an arbitrary point at which $U = 0$. Taking that point to be $x = 0$ (the bottom of the swing), we have $U(x_0) = U(0) = 0$, and

$$U(x) = mgR(1 - \cos\theta) \cong mgR\theta^2/2 = mgx^2/2R,$$

or

$$k = mg/R,$$

and (1.8)

$$F \cong -mgx/R = -mg\theta.$$

Thus $\omega_0^2 \equiv k/m = g/R$, and Equation 1.3 becomes

$$\ddot{x} + gx/R = 0. \tag{1.9}$$

It is easy to verify that this approximate equation of motion for small swings of the pendulum is the same one that follows from the familiar resolution of the vertical force mg into components along the string and along the arc; the latter component is $-mg \sin\theta$, which is, for small θ, the force in Equation 1.8.

A second example of the small-excursion approximation is illustrated in Figure 1.2. Here the mass m is constrained to move (without friction) on a straight rod that lies along the x-axis. Attached to the mass is one end of a spring of unstretched length S_0 and negligible mass; the other end of the spring is attached to a fixed point on the y-axis, $(0, D)$, where $D > S_0$ (the spring is always stretched).

A spring obeying Hooke's law, stretched from its unstretched length S_0 to a greater length S, exerts a restoring force proportional to $S - S_0$:

$$F_s(S) = -k_0(S - S_0), \tag{1.10}$$

where k_0 is known as the spring constant. Thus the energy stored in the spring, equal to the work required to stretch it, is

$$U = -\int_{S_0}^{S} F_s(S')dS' = \tfrac{1}{2}k_0(S - S_0)^2. \tag{1.11}$$

Here we are assuming that there is no dissipation in the stretching and relaxation of the spring, so all the work done in stretching it can be recovered when the spring relaxes. In this case, the work U plays the role of potential energy of the mass m; it can be expressed in terms of x by means of

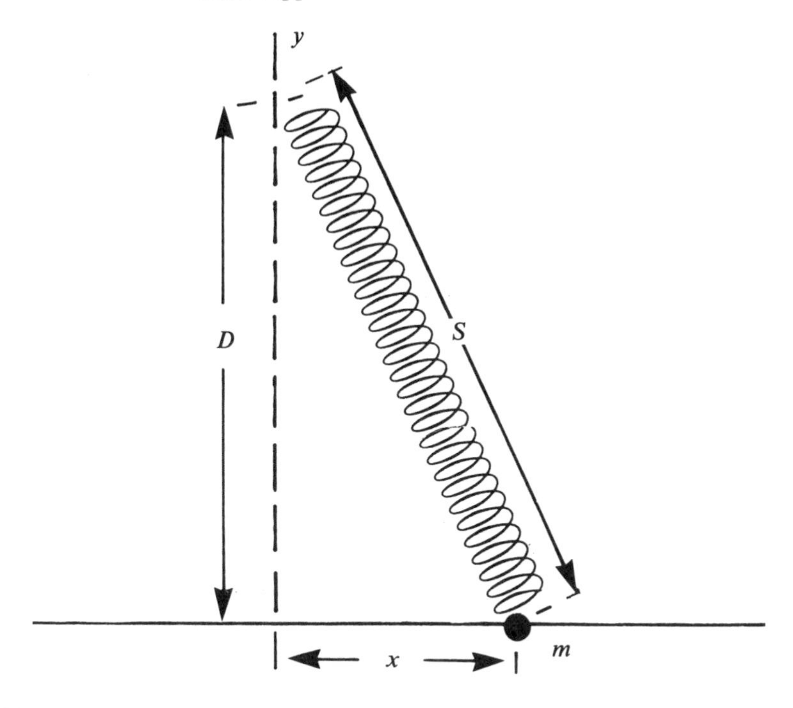

Figure 1.2. Mass oscillating on a line with a restoring force produced by a spring.

$$S^2 = D^2 + x^2, \tag{1.12}$$

whence

$$U = \tfrac{1}{2}k_0(S^2 - 2SS_0 + S_0^2) = \tfrac{1}{2}k_0(D^2 + S_0^2 - 2S_0\sqrt{D^2 + x^2} + x^2). \tag{1.13}$$

Expanding the square root by the binomial theorem for the case $x^2 < D^2$, and dropping all terms past that in x^2, we find that

$$U = U_0 + \tfrac{1}{2}kx^2, \tag{1.14}$$

where

$$U_0 = \tfrac{1}{2}k_0(D - S_0)^2$$
$$k = k_0(1 - S_0/D). \tag{1.15}$$

U_0 is the energy stored in the spring at its minimum length D, which is its length when $x = 0$. The effective spring constant k is related to the actual spring constant k_0 by the factor $1 - S_0/D$, and thus k is always less than k_0, becoming equal to it only in the limit of infinite D (before which any real spring would have been stretched beyond its elastic limit). If D is only slightly greater than S_0, k is very small and there is danger that the Hooke's law approximation, for the force accelerating the mass, may break down. It will be valid only if the maximum value of x^2 is small enough, and the largest allowable value of x^2 depends on how much larger D is than S_0. The condition for the x^2 term in U to be much larger than the next term (in x^4) is

$$x^2/D^2 \ll 4(D/S_0 - 1), \tag{1.16}$$

as one may verify by calculating the x^4 term in the binomial series. If $D \leqslant S_0$, there is no domain of x in which the potential energy is approximated by a quadratic function of x.

Before studying the solutions of Equation 1.3, let us derive the same equation as it applies to a closed electric circuit containing an inductance L in series with a capacitance C, with no resistance (see Figure 1.3).

A condenser with charges Q and $-Q$ on its plates has a capacitance C defined as

$$C = Q/V_c \tag{1.17}$$

where V_c is the potential difference between the plates (potential of the plate bearing charge Q minus that of the other). In a great many cases the capacitance of a given condenser is independent of Q (or of V_c). The potential difference between the plates is always equal to the line integral of the electric field **E** from one plate to the other. If the electric field at each given point between

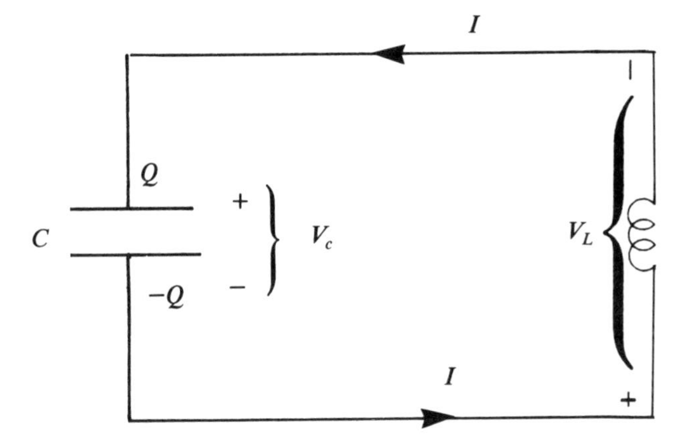

Figure 1.3. LC Circuit.

the plates is proportional to Q, it follows that V_c is proportional to Q, or C is constant. If the medium between the plates is vacuum, the field at a given point on the surface of either plate is proportional to the charge density on the plate at that point (by Gauss' law), and hence to Q; furthermore, the value of E at any other point is proportional to its values at the points on the surfaces of the plates, so the line integral is proportional to Q, or C is constant. If the condenser contains a dielectric, usually the polarization of the dielectric at a given point will be proportional to E at that point, hence to E at the conducting surfaces—so again Q and V_c are proportional to each other. This proportionality will hold true even if the dielectric is anisotropic so that polarization and electric field have different directions, provided only that the two vectors bear a linear homogeneous relationship to each other.

There are two principal cases in which the proportionality of Q and V_c breaks down: the case in which polarization and electric field are related nonlinearly, as when the dielectric saturates; and the case in which the dielectric is not a perfect insulator, but supports a current that may be proportional to V_c—that is, when the condenser in the circuit is properly represented by a capacitance in parallel with a resistance. The diagram in Figure 1.3 implicitly excludes this possibility.

The potential difference across a pure inductance (i.e., one without resistance) is that induced by the changing magnetic flux that links the circuit. This induced potential diference is especially large in the case of a coil, for the flux through the coil when a given current is flowing is roughly proportional to the number of turns and is thus augmented by the coil structure, and the induced voltage caused by a given rate of change of flux is also augmented by the geometry, being again roughly proportional to the number of turns. If a coil has a vacuum (or air) core, the magnetic flux density at a given point is proportional to the current flowing, and its time derivative is (except at very high frequencies) proportional to the time derivative of the current. Thus the time derivative of the flux linking the coil is proportional to the time derivative of the current, and so induces a voltage in the coil that is proportional to the same quantity:

$$V_L = L dI/dt \tag{1.18}$$

where I is the current, and the constant of proportionality L is known as the inductance of the coil. Here I and thus $\dot{I}$ are positive in the direction of the arrow in Figure 1.3, and V_L is positive when the bottom of the coil is at higher potential than the top. Thus Equation 1.18 satisfies Lenz's law.

If the coil has a core that can be magnetized, V_L is still proportional to dI/dt provided that the magnetization at each point is proportional to the magnetic flux density at that point (even though these two vectors may have different directions). Ferromagnetic cores become saturated, however, and exhibit hysteresis. Saturation implies that the relationship between magnetization and flux density is nonlinear, and hysteresis implies that some energy is converted to heat during magnetization and demagnetization of the core or, as we shall see later, that the inductance contains resistance also. This latter case is implicitly ruled out by the absence of resistance in Figure 1.3. When we later study a circuit containing resistance, the effects of hysteresis and of conduction by the dielectric in the capacitor can be approximately taken into account.

The effects of nonlinearity in either the capacitor or in the inductor (i.e., of nonconstancy of C or L) are analogous to the effects of nonlinearity in the mechanical systems that we have discussed previously—in that case equivalent to nonconstancy of the spring constant k. In the electrical case as in the mechanical, such nonlinear effects can be ignored in systems in which the relevant variables are confined to sufficiently small ranges. In most of this book we shall assume in systems of both kinds that we are dealing with small oscillations or variations and hence that nonlinear effects are negligible.

Now having, in Equations 1.17 and 1.18, relations between potential differences and quantities related to charge, we need to use some general principles of circuits in order to get an equation of motion. Such principles are embodied in Kirchhoff's laws:

1. The sum of potential differences taken around any closed path is zero.
2. The total current flowing to any point is zero.

The first of these laws is equivalent to the statement that there is such a quantity as electric potential in a circuit even when potential differences depend on time. The second law follows from conservation of charge and the definition of current as the amount of charge passing a given point per second. In applying the second law to a capacitor, one has to treat its two terminals like a single point without inquiring into what happens inside the device; then the law implies that, at any instant, the current entering one terminal of the capacitor is equal to the current flowing out of the other terminal. One can generalize the law and thus apply it also inside a capacitor, if one generalizes the concept of current to include polarization current and the time derivative of electric-field flux; we shall not need this generalization in what follows.

Kirchhoff's second law, combined with the definition of current as the flow of charge, implies that at any given time the current I has a single value in the circuit, and that its value is

$$I = dQ/dt. \tag{1.19}$$

The first law then requires that the sum of potential differences, taken in a given sense around the circuit (e.g., clockwise) must be zero:

$$Q/C + LdI/dt = 0. \tag{1.20}$$

These two equations imply that

$$Ld^2Q/dt^2 + Q/C = 0,$$

or (1.21)

$$\ddot{Q} + \omega_0^2 Q = 0,$$

where now $\omega_0^2 = 1/LC$.

Equations 1.2 and 1.3 clearly have the same form as Equations 1.21. They can thus be studied together by use of either mechanical or electrical notation and the mechanical case can be translated into the electrical case by means of the following table of correspondences:

$$m \text{ corresponds to } L$$

$$k \text{ corresponds to } 1/C$$

$$x \text{ corresponds to } Q \tag{1.22}$$

$$v = \dot{x} \text{ corresponds to } \dot{Q} = I.$$

Using mechanical notation, let us now study the solutions of Equation 1.3. Because the equation is linear and homogeneous and has constant coefficients, we try the form $x = e^{\lambda t}$. When this form is substituted into the equation, we find that

$$(\lambda^2 + \omega_0^2)e^{\lambda t} = 0,$$

or (1.23)

$$\lambda = \pm i\omega_0.$$

The two roots $\pm i\omega_0$ yield two linearly independent solutions of the equation, and the general solution is a linear combination of the two, with constant coefficients:

$$x(t) = Ae^{i\omega_0 t} + Be^{-i\omega_0 t}. \tag{1.24}$$

This solution can be put in different forms:

$$x(t) = C \cos \omega_0 t + D \sin \omega_0 t, \tag{1.25}$$

$$x(t) = K \cos (\omega_0 t + \theta), \tag{1.26}$$

and

$$x(t) = Re(Je^{i\omega_0 t}). \tag{1.27}$$

The constants A, B, C, D, J, K, θ are in general complex, and all the forms except the last thus represent the general *complex* solution of the equation. If one seeks the general *real* solution, the solution should contain two arbitrary real constants, whereas two complex constants contain four real constants. Thus, to assure that the various forms represent real functions, one must impose two additional conditions on the two complex constants entering each (except Equation 1.27, which is real already). Clearly one must require C, D, K, and θ to be real themselves, and A must be the complex conjugate of B:

$$A = B^*. \tag{1.28}$$

These requirements impose two conditions on the four real constants that are present in each of the forms 1.24–1.26. The form 1.27 contains only two real constants, the real and imaginary parts of J, and x is rendered real by the taking of the real part.

The various forms can all be translated into each other. For example,

$$\left.\begin{aligned} C &= K\cos\theta \\ D &= -K\sin\theta \end{aligned}\right\} \quad \text{and} \quad \left\{\begin{aligned} C &= A + B = A + A^* \\ D &= i(A - B) = i(A - A^*) \end{aligned}\right. \tag{1.29}$$

If J is written in polar form, we have

$$J = |J|e^{i\phi},$$

and

$$\begin{Bmatrix} K = |J| \\ \theta = \phi \end{Bmatrix} \quad \text{or} \quad \begin{Bmatrix} K = -|J| \\ \theta = \phi + \pi \end{Bmatrix}. \tag{1.30}$$

These relations are sufficient to permit one to calculate any set of constants in terms of any other, with little labor.

Thus the general real $x(t)$ depends on two real constants of integration, as befits the general solution of a second-order differential equation. To determine $x(t)$ completely in a given instance one thus needs two additional pieces of information. These can be given in the form of initial conditions that hold at some time such as $t = 0$:

$$\begin{aligned} x(0) &= x_0, \\ \dot{x}(0) &= v_0, \end{aligned} \tag{1.31}$$

where x_0 and v_0 are specified real numbers. Form 1.25 is convenient to use with such initial conditions. One finds that

$$C = x_0$$

and

$$D = v_0/\omega_0. \tag{1.32}$$

Alternatively, one may specify the values of x at two different times:

$$x(t_1) = x_1$$

and

$$x(t_2) = x_2, \tag{1.33}$$

where again x_1 and x_2 are given real numbers. Then

$$x_1 = C \cos \omega_0 t_1 + D \sin \omega_0 t_1$$

$$x_2 = C \cos \omega_0 t_2 + D \sin \omega_0 t_2. \quad (1.34)$$

These equations can be solved for C and D by a standard method such as the method of determinants:

$$C = (x_1 \sin \omega_0 t_2 - x_2 \sin \omega_0 t_1)/\sin \omega_0(t_2 - t_1)$$

and (1.35)

$$D = (x_2 \cos \omega_0 t_1 - x_1 \cos \omega_0 t_2)/\sin \omega_0(t_2 - t_1).$$

These solutions show some of the limitations of boundary conditions given in this form. For example, if $\omega_0(t_2 - t_1)$ is a multiple of π (that is, t_1 and t_2 are separated by an integral number of half cycles of the oscillating function) the denominators in Equation 1.35 vanish and C and D are indeterminate; giving the values of a sinusoidal function at two points that are so related is either redundant or contradictory.

However the constants are determined, $x(t)$ is a sinusoidal function with frequency ω_0. The relation of the velocity $\mathrm{v}(t) \equiv \dot{x}(t)$ to $x(t)$ may be most easily seen by use of the form 1.27. The time derivative of the real part of a complex function is the real part of the time derivative, so

$$\mathrm{v}(t) = Re(i\omega_0 J e^{i\omega_0 t}). \quad (1.36)$$

Thus the amplitude of oscillation of v is that of x multiplied by the frequency ω_0, and the phase of v is related to that of x by the factor i which has been introduced by the differentiation. Because $i = e^{i\pi/2}$, the phase of v exceeds that of x by the angle $\pi/2$, or v leads x by a quarter of a cycle.

Often such quantities as x and v are written as if they were complex, with no indication that one is to take the real part. This abbreviated notation not only saves writing; it also facilitates some calculations (which we shall discuss in Chapter 3). The utility of representing sinusoidal functions of time in complex form is further illustrated through the use of vector diagrams, in which a quantity like $Je^{i\omega_0 t}$ is represented as a vector in the complex plane, the length of the vector being $|J|$ and its angle at time t being $\omega_0 t + \phi$ (recall Equation 1.30). Such a vector rotates counterclockwise around the origin with angular velocity ω_0, and the physical quantity being represented, the projection of the vector on the real axis, oscillates sinusoidally. If $x(t)$ is the projection of a rotating vector, $v(t)$ will be the projection of another which leads the first by 90° as they both rotate. It is often easier to visualize and understand the relationship of two or more such vectors than that of their projections.

To illustrate the use of complex functions and the vectors representing them, let us consider the current I and the potential differences V_L and V_C. Taking Q to be represented by a complex function like that whose real part appears in Equation 1.27, we have

$$I \equiv \dot{Q} = i\omega_0 J e^{i\omega_0 t} \equiv I_0 e^{i\omega_0 t}. \tag{1.37}$$

Then

$$V_L = L\dot{I} = i\omega_0 L I_0 e^{i\omega_0 t} = i\omega_0 L I = L\omega_0 I e^{i\pi/2} \tag{1.38}$$

and

$$V_C = Q/C = I/i\omega_0 C = I e^{-i\pi/2}/\omega_0 C. \tag{1.39}$$

The current is shown in Figure 1.4 as a vector of length $|I_0|$, rotating counterclockwise with angular velocity ω_0. V_L is a vector of length $\omega_0 L$ times the length of I, also rotating as I is, and oriented 90° counterclockwise from (ahead of) I. V_C is a vector of length equal to that of I, divided by $\omega_0 C$, rotating as I is, and oriented 90° clockwise from (behind) I. The lengths of the vectors representing V_L and V_C are equal and the directions are opposite because $V_L = -V_C$.

The possibility of doing calculations with complex functions instead of their real parts depends on the real part of the result of a calculation being the result of doing the calculation on the real part of the complex function. Time differentiation has been seen to have this convenient property. One should not assume, however, that all calculations can be simplified in such a way.

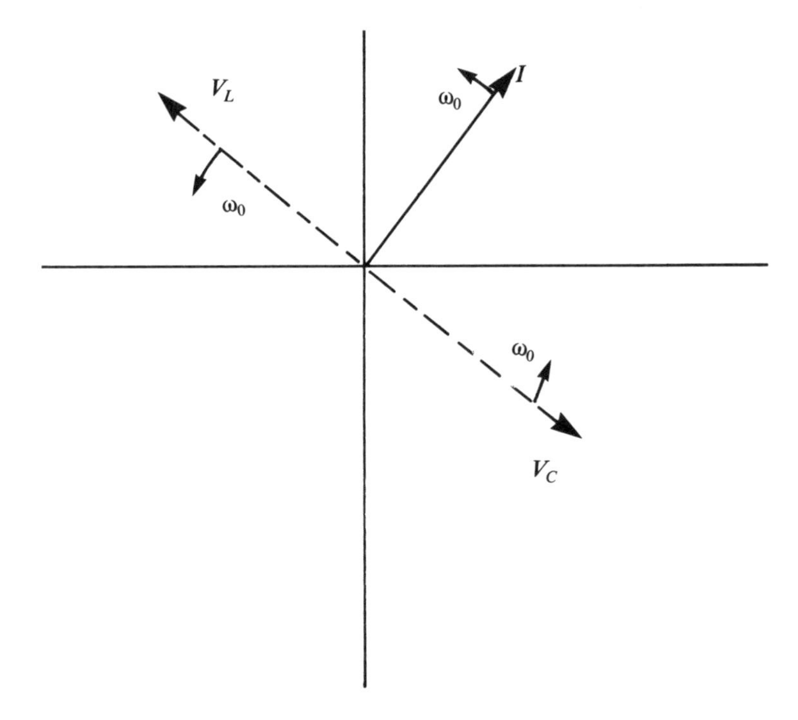

Figure 1.4. Amplitude and phase relationships of voltages and current in an LC circuit.

For example, quadratic functions of x and v must be calculated with real functions x and v, for the real part of such a function is not the same function of the real part of x or v.

This fact affects calculations of energies, for these are quadratic functions of x and v (or of Q and I in the electrical case):

$$T = \tfrac{1}{2}m\text{v}^2 = \tfrac{1}{2}m\dot{x}^2,$$
$$U = \tfrac{1}{2}kx^2 = \tfrac{1}{2}m\omega_0^2x^2, \tag{1.40}$$

where T is kinetic energy and U is, as before, potential energy. The form 1.26 is the most convenient one to use in a calculation of T and U. We find that

$$T = \tfrac{1}{2}m\omega_0^2K^2 \sin^2(\omega_0 t + \theta) = \tfrac{1}{4}m\omega_0^2K^2[1 - \cos(2\omega_0 t + 2\theta)],$$

and (1.41)

$$U = \tfrac{1}{2}m\omega_0^2K^2 \cos^2(\omega_0 t + \theta) = \tfrac{1}{4}m\omega_0^2K^2[1 + \cos(2\omega_0 t + 2\theta)].$$

T and U oscillate with the same amplitude and frequency, but out of phase in such a way that one is large when the other is small. Each varies between zero and $\frac{1}{2}m\omega_0^2K^2 = \frac{1}{2}kK^2$. Clearly their sum is constant:

$$T + U \equiv E = \tfrac{1}{2}m\omega_0^2K^2 = \tfrac{1}{2}kK^2. \tag{1.42}$$

The constant total energy contained in the system periodically changes form between kinetic energy and potential energy. All the energy stored in either form during a quarter cycle returns to the other form during the next quarter cycle; there is no dissipation of energy.

The similarity of the electrical case to the mechanical suggests that the energies in the circuit will behave as do those in the mechanical oscillator. In order to check this indication we recall that, to move a charge dQ through a potential difference V, one must do an amount of work equal to

$$dW = VdQ = VIdt, \tag{1.43}$$

where dt is the time during which the current I is flowing against the potential difference V, transporting a charge dQ in the process. Equation 1.17 gives the relation between Q and V for a capacitor, whence the integral of Equation 1.43 is

$$W_C = \int V_C dQ = \int_0^Q Q'dQ'/C = Q^2/2C = \tfrac{1}{2}V_C Q = \tfrac{1}{2}CV_C^2, \tag{1.44}$$

where the choice of the lower limit of the definite integral corresponds to the arbitrary zero of W_C being set at zero charge.

In the case of the inductance, Equation 1.18 gives the potential difference induced by the changing magnetic flux. This is the potential difference against which one is doing work in moving a positive charge dQ upward through the inductance, for the electric field induced in the coil is pushing positive charge toward the lower end when I is positive. Thus

$$dW_L = L\dot{I}\,dQ = L\dot{I}\,Idt = LIdI, \text{ so } W_L = \int_0^I LI'dI' = \tfrac{1}{2}LI^2. \tag{1.45}$$

Here W_L has been chosen to be zero when $I = 0$.

These two amounts of work, W_L and W_C, can be interpreted respectively as the energies stored in the magnetic field of the inductance and the electric field of the capacitance. The table of correspondences 1.22 between mechanical and electrical quantities shows that

$$T \text{ corresponds to } W_L$$

and (1.46)

$$U \text{ corresponds to } W_C.$$

The previous work on mechanical energy then shows that

$$W_L + W_C = E = \text{constant}. \tag{1.47}$$

The constancy of total energy in both the mechanial and electrical cases indicates that we have somehow built energy conservation into our calculation, and one might wonder whether such a calculation is realistic: Can the energies stored in coils and condensers actually be fully recovered? Are the charging of a condenser and the building-up of current in a coil reversible processes?

In these cases the question reduces to whether VdQ (or $LIdI$, in the case of the inductance) is an exact differential. If these differentials are exact, each is the differential of a function (of Q or of I), and this function, the stored energy, will have a value that depends only on the instantaneous value of Q or I). On the other hand, if the differentials are inexact, such functions of Q and I do not exist and the integral of VdQ or $LIdI$, from an initial value of the independent variable to a new value and back to the initial value, will be nonzero. In such cases there has been a net loss of energy to the coil or the condenser (or conceivably a net gain of energy from such device) in the course of a cyclic change in the rest of the circuit. This sort of effect will occur if, for example, the potential difference across the condenser with given charges on the plates depends on whether the charge is increasing or decreasing. Such behavior is called "hysteresis"; it is more common in iron-core inductances than in capacitances. The possibility of energy gain from hysteresis, as mentioned parenthetically above, cannot be realized; any energy that has been put into one of these devices is recoverable, or has gone into heating the dielectric or the core of the coil. According to the second law of thermodynamics, such heat cannot be reconverted into electric or magnetic energy. Thus hysteresis is a dissipative mechanism; it takes energy from the circuit and does not return it. When we questioned conservation of energy above, we did not mean to question the constancy of *all* the energy in an isolated system, including heat; we referred only to the energy of large-scale, organized motion of charges and currents. *That* energy is not conserved in the presence of hysteresis. It would have been more proper, though perhaps more obscure, to ask whether entropy is constant in the systems that we are studying. As our treatment of the systems presupposes, entropy is constant, no energy is dissipated, and the systems oscillate with constant amplitude.

The foregoing comments about hysteresis and its assumed absence apply also, mutatis mutandis, to our mechanical oscillator. In that case hysteresis might appear in the stretching and relaxation of a spring, leading to irreversible behavior and some generation of heat in the spring. We have assumed that such dissipative effects are absent here too.

We have assumed, also, that our systems are linear, i.e., that their equations of motion are linear differential equations. We shall continue, in most of what follows, to make this assumption, which often amounts to the assumption of small-amplitude oscillations. Nonlinearity is a necessary but not a sufficient condition for hysteresis; linear systems do not exhibit hysteresis, whereas *some* nonlinear systems do. In our subsequent study of linear systems, however, we shall introduce other dissipative effects, which *are* compatible with linearity.

NOTES

The material in this chapter and the next two may be found in many textbooks on mechanics and on electricity. Among such books are:

1. Kip, A. F. *Fundamentals of Electricity and Magnetism*. 2nd ed. New York: McGraw Hill, 1969.
2. Pugh, E. M., and E. W. Pugh. *Principles of Electricity and Magnetism*. 2nd ed. Reading, MA: Addison Wesley, 1970.
3. Symon, K. R. *Mechanics*. 3d ed. Reading, MA: Addison Wesley, 1971.

PROBLEMS

1.1 For the system depicted in Figure 1.2, verify (a) that $U(x)$ contains no cubic term, and (b) that the inequality 1.16 is a correct condition for the quadratic term to be large compared to the quartic term. What does the symmetry of the system imply about the presence of odd powers of x in $U(x)$?

1.2 Assuming a condenser with large parallel plates close together, check the statements after Equation 1.17 pertaining to the constancy of C.

1.3 In the paragraph before Equation 1.18 it is stated that two quantities are "roughly" proportional to the number of turns in a coil. Why "roughly"? What conditions would have to be satisfied to make the rough proportionalities exact?

1.4 An LC circuit has $L = 3 \times 10^{-4}\, h$ and $C = 3 \times 10^{-8} f$. What is the circuit's resonant frequency? If the circuit is oscillating with $10\, j$ energy, what is the peak charge on the capacitor?

1.5 How would the behavior of an LC circuit be changed if C were negative? If L were negative?

1.6 Verify condition 1.28 for $x(t)$ to be real. Using Equations 1.29 and 1.30, calculate A in terms of the magnitude and phase of J. phase of J.

1.7 Verify Equations 1.32 and 1.35.

1.8 What error would one make in calculating T and U for an oscillator with complex functions x and v, and then taking the real parts?

1.9 An argument in the text relates the second law of thermodynamics to the impossibility of gaining energy from an effect such as hysteresis. Complete the argument by filling in the gaps, using an accurate statement of the second law.

1.10 A mass moves without friction on a straight rod. It is attached to two springs with the same spring constant k_0 and the same unstretched length S_0. The other ends of the springs are attached to two fixed points that are both the same distance D from the rod, $D > S_0$, but on opposite sides of the rod, such that the line between the fixed points makes an angle θ with the rod. Calculate the effective spring constant for small oscillations of the mass, assuming that no other force acts on the mass.

1.11 An undamped oscillator of frequency ω_0 is initially at rest. It is to be subjected to two instantaneous impulses of equal magnitude and of the same sign. What should be the time interval between these two impulses if one wishes maximum motion of the oscillator after the second impulse? If one wishes minimum motion?

1.12 An electric charge Q is fixed in place, and a charge $-Q$ slides without friction on a straight line that passes a distance D from the fixed charge. If the moving charge has mass m, what is its frequency of small oscillations about equilibrium?

1.13 A U-tube has cross-sectional area A; its arms are vertical, open at their tops. A mass M of nonviscous liquid fills a length L of the tube. Assuming that the liquid moves in the tube with uniform velocity across any given cross section, determine the frequency of small oscillations of the liquid around its equilibrium configuration. (This problem originated with Isaac Newton.)

1.14 A cylindrical log is weighted at one end so it floats upright in water with a length s submerged. Ignoring the loss of energy to the water, determine the frequency of small vertical oscillations.

1.15 A mass M hanging on a coil spring elongates it by the amount s. While this system is in equilibrium a second mass M is dropped on the first one from a height s, and sticks to it. What are the frequency and the amplitude of the ensuing oscillation?

1.16 How would the answers to the three previous problems be changed if the events occurred on the moon? In an elevator accelerating upward? How would such changes of location affect the period of a pendulum?

2

The Effect of Damping

In the previous chapter we undertook the study of certain systems with linear equations of motion and with no dissipation of energy. In pointing out that hysteresis involved both dissipation and nonlinearity, we remarked that the former can occur without the latter. The most familiar device that dissipates energy in a manner describable by linear differential equations is a resistor in an electric circuit. Accordingly, we shall in this chapter treat electrical systems before mechanical.

Figure 2.1 depicts the circuit that we shall be studying; it differs from the circuit of the last chapter solely through the inclusion of the resistance R. According to Ohm's law, the potential difference across a resistor is

$$V_R = IR, \tag{2.1}$$

where, if the positive directions of I and V_R are as shown in the diagram, R is positive. This expression for potential difference is accurate for a wide range of resistor materials and current densities.

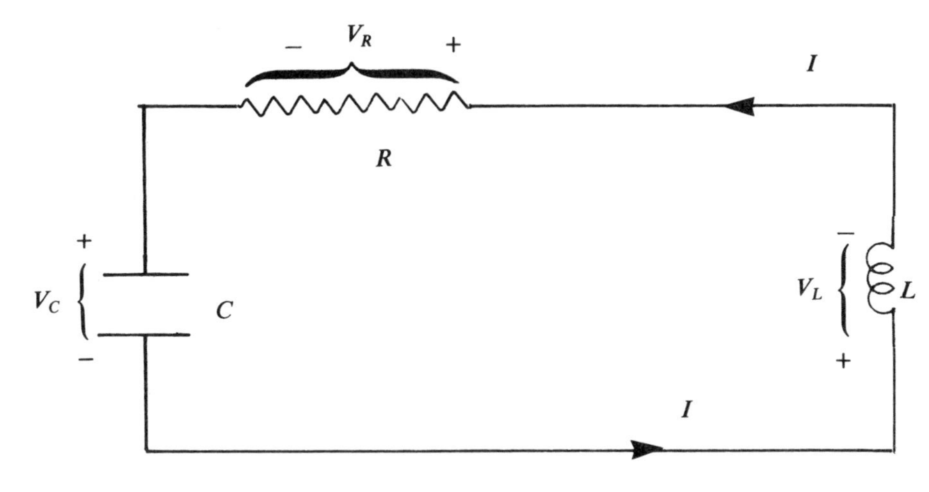

Figure 2.1. An RLC circuit.

Now Kirchhoff's first law implies that

$$L\ddot{Q} + R\dot{Q} + Q/C = 0,$$

or (2.2)

$$\ddot{Q} + 2\gamma\dot{Q} + \omega_0^2 Q = 0.$$

Here ω_0^2 is, as before, $1/LC$, and $\gamma \equiv R/2L$.

Again seeking a solution of the form $e^{\lambda t}$, we find that

$$(\lambda^2 + 2\gamma\lambda + \omega_0^2)e^{\lambda t} = 0 \quad (2.3)$$

which quadratic equation has two roots,

$$\lambda = -\gamma \pm \sqrt{\gamma^2 - \omega_0^2}. \quad (2.4)$$

Again, as must happen in a second-order equation, the two roots yield two linearly independent solutions, and the general solution is a linear combination of the two with arbitrary constant coefficients. The only exceptional case in which the general solution takes a different form is that in which the two roots of Equation 2.4 coincide; in that case it can be verified that the following expression satisfies the equation:

$$Q(t) = (A + Bt)e^{-\gamma t}. \tag{2.5}$$

This case ($\omega_0 = \gamma$) is known as the case of *critical damping;* it is the boundary between the more usual cases of *underdamping* ($\omega_0 > \gamma$) and *overdamping* ($\omega_0 < \gamma$).

In the cases of overdamping and critical damping the general solution of Equation 2.2 contains no complex or oscillatory functions, so there is no occasion to present the solution in alternative forms like those of the last chapter, or to represent real solutions by means of complex functions. The general solution for the case of critical damping is given above; that for the overdamped circuit is

$$Q(t) = e^{-\gamma t}(Ae^{\mu t} + Be^{-\mu t}), \tag{2.6}$$

where

$$\mu = \sqrt{\gamma^2 - \omega_0^2}.$$

In both these ases the constants A and B are real, and can be determined by initial (or other equivalent) conditions.

We shall devote more attention to the case of underdamping. We define

$$\omega' = \sqrt{\omega_0^2 - \gamma^2} \tag{2.7}$$

and write solutions in four forms that correspond to those in Equations 1.24–1.27:

$$Q(t) = e^{-\gamma t}(Ae^{i\omega' t} + Be^{-i\omega' t}), \tag{2.8}$$

$$Q(t) = e^{-\gamma t}(C \cos \omega' t + D \sin \omega' t), \tag{2.9}$$

$$Q(t) = Ke^{-\gamma t} \cos (\omega' t + \theta), \tag{2.10}$$

$$Q(t) = Re(Je^{(i\omega' - \gamma)t}) = Re(Je^{i\omega' t})e^{-\gamma t} \tag{2.11}$$

Except for the common factor $e^{-\gamma t}$, these forms are similar to those of the previous chapter; their conditions of reality and the relations among the constants in the different forms are similar, also, to the earlier ones.

Initial conditions are conveniently applied to form 2.9. Letting

$$Q(0) = Q_0$$

and (2.12)

$$\dot{Q}(0) \equiv I(0) = I_0,$$

we find that

$$C = Q_0$$

and (2.13)

$$D = (I_0 + \gamma Q_0)/\omega'.$$

We leave as a problem for the reader the application of conditions on Q at two different times, as in Equations 1.33–1.35.

To investigate the relation of current to charge we use form 2.11. Taking the time derivative of the complex function and then taking the real part of the time derivative, we find that

$$I(t) \equiv \dot{Q}(t) = Re[(i\omega' - \gamma)Je^{(i\omega' - \gamma)t}]. \tag{2.14}$$

Similarly,

$$\dot{I}(t) = Re[(i\omega' - \gamma)^2 Je^{(i\omega' - \gamma)t}]. \tag{2.15}$$

Letting

$$i\omega' - \gamma = \omega_0 e^{i\psi}, \; \omega_0 = \sqrt{\omega'^2 + \gamma^2}, \tag{2.16}$$

we see that the periodic factor in I leads that in Q, and that in $\dot{I}$ leads that in I, by the angle ψ, which lies between 90° and 180°. The presence of resistance in the circuit does not alter the potential differences V_C and V_L, for we define these quantities to be applicable to resistance-free ideal circuit elements; they are not taken to be the potential differences across actual articles of hardware which do in general contain part of the circuit's resistance. Thus,

$$V_C = Q/C = I/C(i\omega' - \gamma),$$

and (2.17)

$$V_L = L\dot{I} = L(i\omega' - \gamma)I.$$

Also

$$V_R = IR.$$

Here, for brevity, we are writing complex quantities and not indicating that real parts are to be taken. By using the definitions of γ and ω' in terms of R, L, and C, and the definition of ω_0 from Chapter 1, one can verify that, as Kirchhoff's first law requires, the sum of the potential differences in Equation 2.17 is zero—indeed, the complex sum is zero although only its real part has to be. The vector diagram in Figure 2.2 depicts the amplitude and phase relationships in the circuit.

In interpreting Figure 2.2 one must keep in mind that all the vectors are not only rotating with the common angular velocity ω'; they are also continuously diminishing in length by the factor $e^{-\gamma t}$. The circuit is not oscillating, like the undamped one, with constant amplitude, in the manner of simple harmonic motion; it is performing damped oscillations, with amplitude asymptotically approaching zero. A useful measure of the rate of damping is the so-called *logarithmic decrement,* defined as the reduction of the natural logarithm of Q in the course of one cycle of the periodic function that enters Q. This reduction is due to the factor $e^{-\gamma t}$ as this factor diminishes over the period $2\pi/\omega'$ of a cycle. The logarithmic decrement is readily seen to be

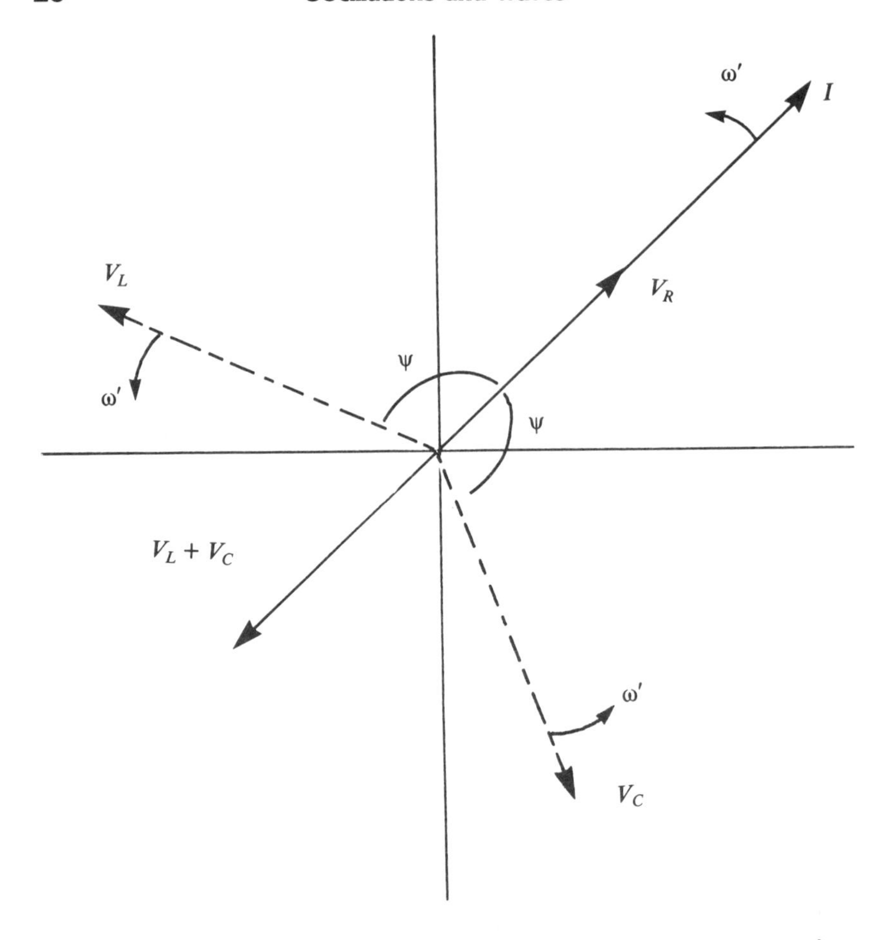

Figure 2.2. Amplitude and phase relationships of voltages and current in an RLC circuit.

$$\delta = \ln \frac{Q(t)}{Q\left(t + \dfrac{2\pi}{\omega'}\right)} = 2\pi\gamma/\omega' = 2\pi/(4L/R^2C - 1)^{1/2}. \qquad (2.18)$$

The fact that, as time passes, the amplitude of oscillation diminishes, suggests that the circuit is somehow losing energy. Having defined V_C and V_L to be the same functions of Q and its derivatives as in the undamped case, we are again led by the same reasoning to the same expressions for W_L and W_C:

$$W_L = \tfrac{1}{2}LI^2,$$

and (2.19)

$$W_C = \tfrac{1}{2}Q^2/C.$$

These quantities, being the integrals of exact differentials, cannot account for any net loss of energy. But an attempt to define a similar quantity W_R fails:

$$dW_R = V_R dQ = R\dot{Q}\,dQ = R\dot{Q}^2 dt = RI^2 dt \qquad (2.20)$$

is not an exact differential; i.e., it is not the differential of any function—there is no such function as W_R, "the energy stored in the resistance." The integral of dW_R toward increasing t is always positive (or zero when I is zero); when any current is flowing, work is being done on the resistance, which never does any work in return on the rest of the circuit. Thus resistance is a purely dissipative mechanism; it dissipates all the energy that enters it, and returns none.

The purely dissipative resistance and the purely *reactive* (energy-storing and energy-returning) inductance and capacitance are idealizations. Actual items of hardware are seldom so "pure": resistors have some inductance, capacitors have some resistance (except in superconducting circuits,etc.). In studying a whole circuit rather than its elements in isolation, however, one can properly lump together the resistance of all the objects in the circuit, indicate it all by a single wavy segment and label it "R," and treat L and C likewise. We shall henceforth assume, as we have implicitly done already, that the symbols R, L, and C are to be so interpreted.

In order to study the storage and dissipation of energy in detail, we use form 2.10 for Q in writing explicitly time-dependent W_C and W_L. Thus

$$Q = Ke^{-\gamma t}\cos(\omega' t + \theta)$$

and (2.21)

$$\dot{Q} = I = \omega_0 K e^{-\gamma t}\cos(\omega' t + \theta + \psi),$$

where ψ is the angle defined in Equation 2.16:

$$\cos\psi = -\gamma/\omega_0$$
$$\sin\psi = \omega'/\omega_0. \tag{2.22}$$

Then

$$W_C = Q^2/2C = K^2 e^{-2\gamma t}\cos^2(\omega' t + \theta)/2C$$

and (2.23)

$$W_L = LI^2/2 = K^2 e^{-2\gamma t}\cos^2(\omega' t + \theta + \psi)/2C.$$

where the definition of ω_0 has been used in the latter equation. Note that ψ is an angle somewhat greater than 90°—closer to 90° when γ/ω_0 is small—so the arguments of the two $\cos^2$ functions are somewhat more than 90° apart; if their difference were exactly 90°, the functions would have a constant sum, and the total energy would change only by virtue of the exponential damping factor. In fact the sum $W_L + W_C$ depends on time through the exponential factor and through a small oscillation coming from the $\cos^2$ functions. With the aid of trigonometric identities, the energy can be written as

$$E = W_L + W_C = K^2 e^{-2\gamma t}[1 + \cos\psi\cos(2\omega' t + 2\theta + \psi)]/2C, \tag{2.24}$$

which expression clearly reveals the two types of time dependence. If γ/ω_0 is very small, one can fruitfully define $\bar{E}$, the energy averaged over one cycle of the oscillatory function with the exponential treated as constant during the averaging:

$$\bar{E} = K^2 e^{-2\gamma t}/2C. \tag{2.25}$$

In fact, regardless of how close to 1 the ratio γ/ω_0 may be, E is a monotonically decreasing function of time. Its time derivative can be put in the form

$$\dot{E} = -\gamma K^2 e^{-2\gamma t}[1 + \cos(2\omega' t + 2\theta + 2\psi)]/C, \quad (2.26)$$

which is always negative except at two instants during each cycle when it is zero (when I is zero). This time derivative, averaged over a cycle of the oscillatory function, is identical to the time derivative of $\bar{E}$:

$$\dot{\bar{E}} = \bar{\dot{E}} = -\gamma K^2 e^{-2\gamma t}/C. \quad (2.27)$$

So when $\gamma \neq 0$ the energy stored in L and C decreases with time; the proportionality of γ to R suggests that, as one might expect, the missing energy has been dissipated in the resistance. In fact, $-\dot{E}$ at any given instant should equal the rate $dW_R/dt = RI^2$ at which work is being done on the resistance. An explicit calculation verifies that this is the case:

$$-\dot{E} = dW_R/dt = I^2 R. \quad (2.28)$$

Equation 2.23 reveals that, if $e^{-2\gamma t}$ is regarded as approximately constant during a cycle of the $\cos^2$ functions entering W_L and W_C, these functions have the same peak value; the peak (or the average) energies stored in L and in C are equal. Thus either of these average values provides a measure of the amount of energy storage that takes place in the circuit. It is sometimes useful to compare this amount,

$$\bar{W} = K^2 e^{-2\gamma t}/4C \quad (2.29)$$

with the amount of energy dissipated in one cycle, that is, $-\dot{\bar{E}}$ multiplied by the duration of one cycle, $2\pi/\omega'$. The ratio of the first of these quantities to the second is sometimes called the ratio of reactive to active power; it is

$$\frac{\bar{W}}{\frac{-2\pi}{\omega'}\dot{\bar{E}}} = \frac{\omega'}{8\pi\gamma} = \frac{1}{4\delta} \equiv \frac{Q}{4\pi} = \frac{\omega' L}{4\pi R}. \quad (2.30)$$

As indicated, this ratio is inversely proportional to the logarithmic decrement δ as given in Equation 2.18. The quantity Q defined in Equation 2.30 is commonly used as a measure of the smallness of dissipation in a circuit, the symbol was originally an abbreviation of "quality," nondissipative circuits being considered good ones for reasons that we shall see shortly.

The foregoing treatment of an electric circuit can be applied, with appropriate changes, to a mechanical damped oscillator. Aside from substitution of mechanical quantities for the corresponding electrical ones, the principal alterations involve energies and the nature of the dissipative mechanism. W_L and W_C are respectively replaced by the kinetic energy T and the potential energy U. These quantities are thus taken to be definable as functions of x and $\dot{x}$. This assumption is true of T, by definition, but potential energy is sometimes said not to exist in a nonconservative system in which a dissipative force is acting. We are, instead, defining U to be the potential-energy function from which the *conservative part* of the force can be derived as $-dU/dx$. Then, since there is also a force that cannot be so expressed, $T + U$ is not conserved—as $W_L + W_C$ is not.

The dissipative mechanism in mechanical systems is commonly called "friction," although it may not involve solid objects sliding on other solid objects. In the systems depicted in Figures 1.1 and 1.2, energy may be lost to the ambient air by action of the air's viscosity; or to heating of the pendulum's suspension at the ceiling through sliding friction or through bending of wires or cables; or to hysteresis in the spring in Figure 1.2. We have already commented on the nonlinearity of hysteresis; probably none of the other mechanisms of energy loss is strictly linear, either, although at low speeds the action of fluid viscosity is nearly so. Thus, although the representation of frictional force as

$$F_R = -R\dot{x} \tag{2.31}$$

may often be an adequate approximation, it is seldom so accurate as Ohm's law. Correspondingly, then, our treatment of damped systems is likely to be more accurate as applied to electrical systems than to mechanical ones. Indeed, even at frequencies high enough to produce appreciable energy loss from a circuit by radiation—an effect that we have not explicitly considered in our work—the effect on the circuit of this radiation can be accurately reproduced by introduction into the circuit of an additional resistance, known as "radiation resistance," which obeys Ohm's law.

NOTES

See references such as:

1. Kip, A. F. *Fundamentals of Electricity and Magnetism.* 2d ed. New York: McGraw Hill, 1969.
2. Pugh, E. M. and E. W. Pugh. *Principles of Electricity and Magnetism.* 2d ed. Reading, MA: Addison Wesley, 1970.
3. Symon, K. R. *Mechanics.* 3d ed. Reading, MA: Addison Wesley, 1971.

PROBLEMS

2.1 Explicitly verify that the solutions 2.5–2.11 satisfy Equation 2.2 in the appropriate cases.

2.2 How would the behavior of an *RLC* circuit be changed if R were negative?

2.3 Evaluate the constants in the underdamped solution so as to fit specified values of Q at two times t_1 and t_2. Are there any choices of times and corresponding Q values that yield contradictory or indeterminate results? Explain.

2.4 Verify that the angle ψ defined in Equation 2.16 is in the second quadrant, and that it appears correctly in Equations 2.22, 2.24, 2.26.

2.5 Fill in the gaps in the reasoning related to energy and its dissipation. In particular, verify Equations 2.24, 2.26, and 2.28.

2.6 Derive Equation 2.30.

2.7 Referring to Problem 1.11, answer the questions for the case of an underdamped oscillator.

2.8 Calculate the motion of an underdamped oscillator for all $t > 0$ in each of the following cases: a) $F(t) = A$ for $t < 0$, $F(t) = 0$ for $t > 0$; b) $F(t) = 0$ for $t < 0$, $F(t) = A$ for $t > 0$.

2.9 Repeat Problem 2.7 for an overdamped oscillator.

2.10 In Problem 1.13, if (on account of energy loss to viscosity) the peak kinetic energy of the oscillating liquid is reduced by a fraction f in each cycle of oscillation, by what fraction will the period of oscillations be changed as a result? Assume $f \ll 1$.

3

Sinusoidally Driven Oscillators and Circuit Loops

Having now studied the free oscillations of harmonic oscillators and RLC series circuits, let us consider the behavior of these systems when they are acted on by external influences. Using mechanical notation, we shall discuss the solutions of the equation of motion when an additional force term is present, given explicitly as a function of t. Then we shall specialize to the case in which this term is a sinusoidal function of t, and, in later chapters, shall generalize the time dependence of the applied force.

If an applied force $F(t)$ is acting, the equation of motion is

$$m\ddot{x} = -R\dot{x} - kx + F(t)$$

or (3.1)

$$\ddot{x} + 2\gamma\dot{x} + \omega_0^2 x = F(t)/m \equiv f(t).$$

Here the symbols are as previously defined, the symbol R being used for mechanical as well as electrical resistance; $f(t)$ is defined as the applied force $F(t)$ divided by the mass.

Equations like 3.1 are known as linear inhomogeneous equations (with constant coefficients). The corresponding homogeneous equation has $f(t) = 0$. Its general solution can be translated into mechanical notation from equation 2.5 or equation 2.6 or equations 2.8–2.11. These general solutions must each contain two unspecified real constants which we shall here designate A and B although they were called various things in equations 2.8–2.11. We shall write the general undriven $x(t)$ as $x_H(t; A, B)$, making the dependence of x on A and B explicit.

Now, letting $x_I(t)$ be any particular solution of the whole inhomogeneous Equation 3.1, we can use a theorem which states that the general solution $x(t; A, B)$ can be written as the sum of the two functions mentioned above:

$$x(t; A, B) = x_H(t; A, B) + x_I(t). \tag{3.2}$$

The linearity of the equation guarantees that, if this sum is substituted for x in the equation, the right side will be the sum of the contributions of the two terms on the right side of Equation 3.2, that is, zero plus $f(t)$. Thus Equation 3.2 satisfies Equation 3.1. Furthermore, the sum of x_H and x_I depends on the two constants A and B; Equation 3.2 gives a two-parameter family of solutions of Equation 3.1 and, x_I being the same in all members of this family, x inherits the property of x_H, that at most two members of this two-parameter family are linearly independent. So the x in Equation 3.2 meets the tests of a general solution of Equation 3.1.

It may still be puzzling that this $x(t; A, B)$ is the general solution of Equation 3.1 when x_I was taken to be *any* solution of Equation 3.1, and x_H does not by itself satisfy Equation 3.1 at all. In fact, the difference of any two solutions of Equation 3.1 is a solution of the homogeneous equation, i.e., a special case of x_H with particular A and B; this fact can be verified by direct substitutions of such a difference into Equation 3.1. Thus, a change from x_I to a different particular solution x_I' induces a change of A and B to new values A' and B', these changes all together leaving x invariant:

$$x_H(t; A, B) + x_I(t) = x_H(t; A', B') + x_{I'}(t). \qquad (3.3)$$

So a different choice of $x_I(t)$ leaves the two-parameter family of solutions of Equation 3.1 invariant. In this sense, one $x_I(t)$ is as good as another. We shall find that, in a wide variety of cases, there is one particular solution $x_I(t)$ that is zero when $f(t)$ is zero, such a solution can reasonably be called the *response* of the system to the force $F(t) = mf(t)$. When we know such an $x_I(t)$ we shall find it a convenient one to use.

Now let us specialize the force to one with a sinusoidal time dependence of unspecified real frequency ω:

$$\begin{aligned} F(t) = mf(t) &= Re(F_0 e^{i\omega t}) \\ &= |F_0| \cos(\omega t + \eta), \end{aligned} \qquad (3.4)$$

where

$$F_0 = |F_0| e^{i\eta}.$$

Since Equation 3.1 is linear and has real coefficients, we can use complex functions $f(t)$ in it, thus getting complex $x(t)$, knowing that the real part of x will satisfy the equation with the real part of f in it. Thus we shall use complex functions $f(t)$ of the form $F_0 e^{i\omega t}/m$, with real ω and complex F_0, and calculate complex solutions x, knowing that the real part of x will satsify the equation with the real part of f on the right. To save writing, we shall omit the designation of real parts and simply take it for granted. Later we shall encounter cases in which all imaginary parts vanish despite the use of complex notation; then it will not matter whether we take real parts, for the solutions will be real in any case.

Thus we seek complex solutions of

$$\ddot{x} + 2\gamma\dot{x} + \omega_0^2 x = F_0 e^{i\omega t}/m. \qquad (3.5)$$

We shall take x_H to be, in the underdamped case,

$$x_H(t) = Je^{(i\omega' - \gamma)t}, \qquad (3.6)$$

where J is complex, as in Equation 2.11. The two constants in the real part of x are the real and imaginary parts, or the magnitude and phase, of J; although this x is not a general complex solution, it yields a general real solution and is thus sufficiently general for our purposes.

Now we seek a particular complex solution of Equation 3.5; the following trial form is strongly suggested:

$$x_I(t) = s_0 e^{i\omega t}, \tag{3.7}$$

where s_0 is a complex constant. On substituting from Equation 3.7 into Equation 3.5, we find that

$$(-\omega^2 + 2i\omega\gamma + \omega_0^2)s_0 e^{i\omega t} = F_0 e^{i\omega t}/m \tag{3.8}$$

or

$$s_0 = F_0/m(-\omega^2 + 2i\gamma\omega + \omega_0^2).$$

Thus the complete complex solution is

$$x(t) = Je^{(i\omega' - \gamma)t} + \frac{F_0 e^{i\omega t}}{m(-\omega^2 + 2i\gamma\omega + \omega_0{}^2)}. \tag{3.9}$$

Both the constants of integration are in the first term, contained in J; the second term is completely specified. The decision to seek a solution of the form $e^{i\omega t}$ determined $x_I(t)$ uniquely. This x_I is the only solution with this time dependence.

Note that this x_I also has the property mentioned above, of vanishing when $f(t)$ (or F_0) vanishes. Thus it would seem to be the response of the system to the specified force. One has to be careful in making such as assertion, however, as we shall see when we contemplate the use of initial conditions.

The complex expression for v is the time derivative of Equation 3.9;

$$v(t) = (i\omega' - \gamma)Je^{(i\omega' - \gamma)t} + \frac{i\omega F_0 e^{i\omega t}}{m(-\omega^2 + 2i\gamma\omega + \omega_0^2)}. \tag{3.10}$$

Recalling that

$$i\omega' - \gamma = \omega_0 e^{i\psi} \tag{3.11}$$

and setting

$$Z = 2m\gamma + i\omega m + m\omega_0^2/i\omega = R + i(\omega m - k/\omega) \equiv R + iX, \tag{3.12}$$

we have

$$v(t) = \omega_0 J e^{-\gamma t + i(\omega' t + \psi)} + F_0 e^{i\omega t}/Z. \tag{3.13}$$

Also, from Equation 3.9,

$$x(t) = J e^{-\gamma t + i\omega' t} + F_0 e^{i\omega t}/i\omega Z. \tag{3.14}$$

In each of these expressions, the first term contains the factor $e^{-\gamma t}$ which goes asymptotically to zero with increasing t; these terms are thus called "transient" terms. The transient terms contain the adjustable constants and therefore depend on initial conditions. The second term in each equation oscillates with constant amplitude, at the frequency of the applied force; these terms are called "steady-state" terms. They do not depend on initial conditions, being determined completely by the amplitude, phase, and frequency of the applied force and, by way of the *impedance* Z, by the properties of the driven system.

In order to see how the transient terms are influenced by initial conditions, let us again denote initial position by x_0 and initial velocity by v_0. These quantities are real, and must thus be equated to real parts of the expressions 3.13 and 3.14 with $t = 0$. This procedure leads to straightforward but rather inconvenient equations for the magnitude and phase of J. We shall, instead, let x' and v' be the imaginary parts, at $t = 0$, of the complex x and v:

$$x(0) = x_0 + ix', \text{ and } v(0) = v_0 + iv',$$

and (3.15)

$$x^*(0) = x_0 - ix', \text{ and } v^*(0) = v_0 - iv'.$$

Here x^* and v^* are the complex conjugates of the complex x and v, which are to be equated to the complex conjugates of expressions 13 and 14 with $t = 0$. The two complex equations and their complex conjugates are thus four simultaneous linear equations in which x', v', J, and J^* can be regarded as unknowns. In fact, it is necessary to solve only for J, since J^* is known when J is, and x' and v' are of no interest. The solution is

$$J = x_0(1 - i\,\gamma/\omega') - iv_0/\omega' + F_0\,[\gamma + i(\omega + \omega')/2Z\,\omega\omega' + F_0^*\,[-\gamma + i(\omega - \omega')]/2Z^*\omega\omega'. \tag{3.16}$$

This solution determines the transient terms of $x(t)$ and $v(t)$. Evidently the magnitude and phase of the transient that follows from given initial conditions depend on what sinusoidal force is acting after $t = 0$. The presence of F_0 in the expression for J indicates that the transient is, in part, a response of the system to the applied force. The x_0 and v_0 terms, on the other hand, survive even when $F_0 = 0$, and thus represent the part of the motion that would occur in an undriven oscillator subjected to the given initial conditions. Such initial motion must have been caused by forces that acted at earlier times.

If no such earlier forces have acted, or if they have acted so long ago that the resulting motion has, by $t = 0$, been damped to negligible amplitude, it follows that x_0 and v_0 are both zero. Then J contains only the terms proportional to F_0 and F_0^*, and each term in $x(t)$ and $v(t)$ is linear in F_0. It is appropriate, then, to speak of x and v as the *response* of the oscillator to the given force, but the force must have been zero for negative t since it produced no motion before $t = 0$:

$$\begin{aligned} F(t) &= 0 \text{ for } t < 0 \\ &= F_0 e^{i\omega t} \text{ for } t > 0. \end{aligned} \tag{3.17}$$

In all cases, Equations 3.13 and 3.14 reduce to their steady-state terms at sufficiently large positive t. Or, if $F_0 e^{i\omega t}$, and no other force, has been acting since $t = -\infty$, the steady-state terms are the only ones present even at $t = 0$. Thus it is appropriate to call the steady-state $x(t)$ and $v(t)$ the *response* of the system to the force $F_0e^{i\omega t}$, it being understood that this force and no other has been acting since $t = -\infty$. When we wish to describe an applied force that is represented by different analytic functions of t at different times, we shall (as in Equation 3.17) indicate the fact in our description. When we simply write $F_0e^{i\omega t}$, with no qualifications, we shall mean that the force is represented by the given function of t at all times from $-\infty$ to $+\infty$. The response to such a force is the steady-state part of the solution given in Equations 3.13 and 3.14:

$$x(t) = F_0e^{i\omega t}/i\omega Z;$$
$$v(t) = F_0e^{i\omega t}/Z; \tag{3.18}$$

Z is defined in equation 3.12. These are sinusoidal functions, of frequency equal to that of the applied force; the amplitudes (of the real parts) are

$$|x| = |F_0|/|Z|\omega;$$
$$|v| = |Z_0|/|Z|. \tag{3.19}$$

These amplitudes exhibit a property known as "resonance"; they are particularly large if the frequency ω of the driving force is close to the natural frequency of the oscillator, ω_0. This behavior is particularly perspicuous in the case of v (or current, I, in the electrical case), being entirely dependent on the properties of the impedance, Z. Equation 3.12 gives

$$Z = R + i(\omega m - k/\omega) = R + iX, \tag{3.12$'$}$$

where R, the resistance, is as previously defined, and X, the *reactance,* is the imaginary part of Z:

$$X = \omega m - k/\omega. \tag{3.20}$$

Instead of representing the complex number Z in terms of real and imaginary parts, we can write

$$Z = |Z| e^{i\rho},$$

where

$$|Z| = (R^2 + X^2)^{1/2} = [R^2 + (\omega m - k/\omega)^2]^{1/2}, \tag{3.21}$$

and

$$\tan\rho = X/R = (\omega m - k/\omega)/R.$$

Clearly, if all parameters except frequency are held constant, and the frequency is varied from 0 to ∞, the amplitude of velocity is proportional to $1/|Z|$, being zero at $\omega = 0$, maximum at $\omega = (k/m)^{1/2} = \omega_0$, and zero again at $\omega = \infty$. The maximum value of $|v|$, at ω_0, is $|F_0|/R$. One can get a measure of the width of the peak near ω_0 where $|v|$ is large by seeking the frequencies ω_+, ω_-, where $|v|$ is $|v|_{max}/2^{\frac{1}{2}}$; at these frequencies,

$$X^2 = R^2 \text{ or } \omega m - k/\omega = \pm R,$$

whence

$$\omega^2 \pm 2\gamma\omega - \omega_0^2 = 0, \tag{3.22}$$

so

$$\omega = \pm\gamma \pm \sqrt{\omega_0^2 + \gamma^2}.$$

Of these four values of ω, two are positive and thus physically acceptable:

$$\omega_\pm = (\gamma^2 + \omega_0^2)^{1/2} \pm \gamma,$$

and (3.23)

$$\omega_\pm \cong \omega_0 \pm \gamma,$$

if $\gamma \ll \omega_0$. So in the case of very small damping, $R \ll (km)^{1/2}$, the width of the resonant peak in the graph of $|v|$ against ω is approximately 2γ; the height of the peak is $|F_0|/R = |F_0|/2L\gamma$.

Another aspect of the phenomenon of resonance is the behavior of the phase difference ρ between F_0 and v (the phase angle of Z). At the resonant frequency ω_0, where $X = 0$, this angle is zero also; v and F_0 are in phase with each other at resonance. At $\omega = 0, X = -\infty$, and $\rho = -\pi/2$; at $\omega = \infty, X = +\infty$, and $\rho = +\pi/2$. To get a measure of the range of frequencies in which most of this change of ρ is taking place, we can ask at what frequencies $\tan\rho = \pm 1$. These turn out to be the same frequencies $\omega_\pm$ at which $|v|$ equals its peak value over $\sqrt{2}$, so the phase angle changes most rapidly in the neighborhood of resonance in just those cases in which the resonant peak of $|v|$ is sharpest. Letting $\Delta\omega$ be the width of the range in which $|v|$ (i.e., $|Z|$) and ρ change rapidly, we have $\Delta\omega = \omega_+ - \omega_- = 2\gamma$, and

$$\Delta\omega/\omega_0 = 2\gamma/\omega_0 = \tfrac{1}{2}R(km)^{1/2} \cong (1/Q), \qquad (3.24)$$

where Q is the same one introduced in Equation 2.30. It is usually defined in the electrical case as the ratio, at resonance, of either inductive or capacitative reactance to resistance. Thus we see that an oscillator with a narrow resonance is also one whose free vibrations are damped slowly; the fractional amount of damping per cycle is of the same order of magnitude as the fractional width of the resonance—and also, as we shall see later, the fractional width of the frequency spectrum of the oscillator's damped motion. Figure 3.1 shows $1/|Z|$ and ρ plotted against ω in two cases that have values of R different by a factor 2.

In discussing energy and power in the steady-state motion of the oscillator, we must work with the real expressions for $x(t)$ and $v(t)$:

$$\begin{aligned} x(t) &= |F_0| \cos(\omega t + \eta - \rho - \pi/2)/\omega |Z| \\ &= |F_0| \sin(\omega t + \eta - \rho)/\omega |Z|, \end{aligned} \qquad (3.25)$$

and

$$v(t) = |F_0| \cos(\omega t + \eta - \rho)/|Z|.$$

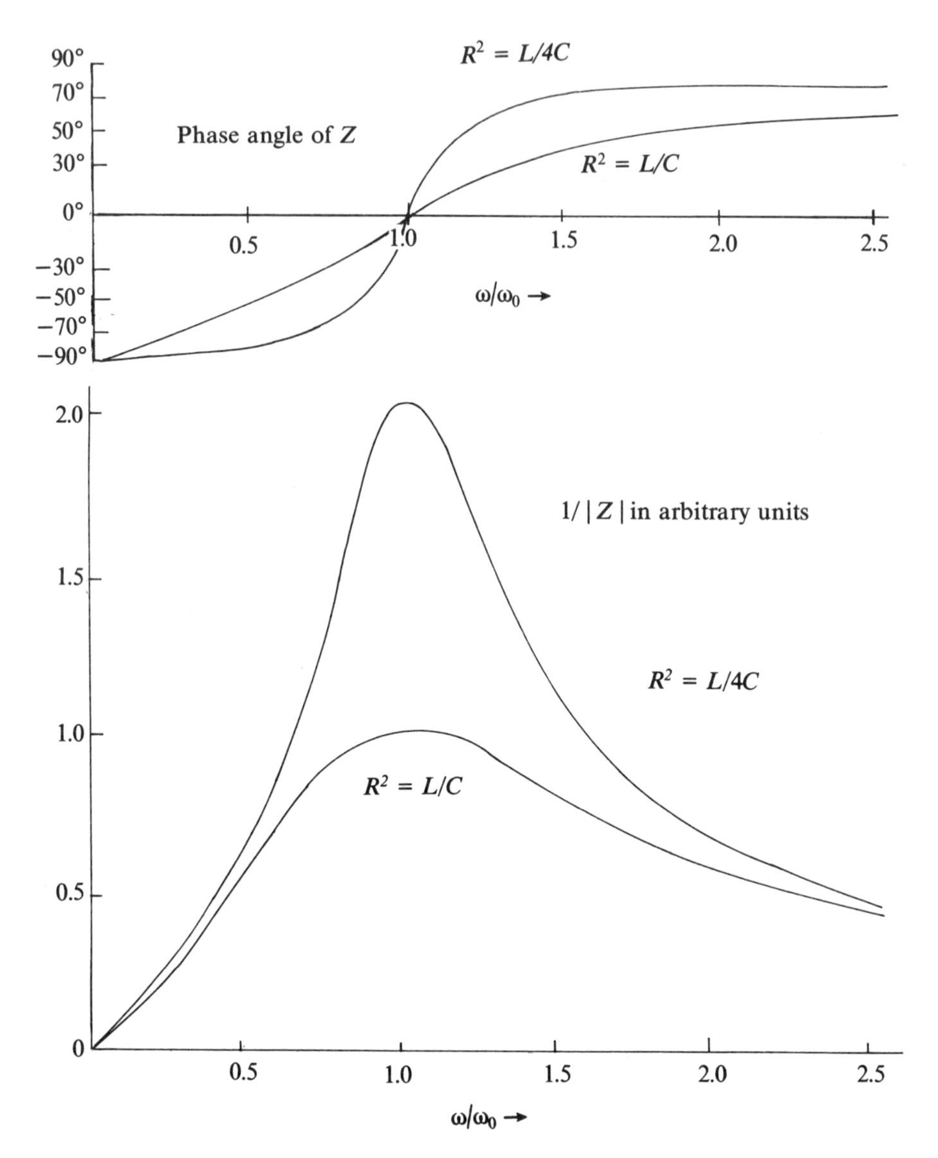

Figure 3.1. Phase and amplitude dependence of impedance on frequency of driving voltage, for two different cases.

Thus kinetic energy is

$$T = \tfrac{1}{2}mv^2 = \tfrac{1}{2}m\,|\,F_0\,|^2\cos^2(\omega t + \eta - \rho)/\,|\,Z\,|^2, \qquad (3.26)$$

and potential energy is

$$U = \tfrac{1}{2}\,kx^2 = \tfrac{1}{2}k|\,F_0\,|^2\sin^2(\omega t + \eta - \rho)/\,|\,Z\,|^2\omega^2$$
$$= \tfrac{1}{2}\,\frac{{\omega_0}^2}{\omega^2}\,m|\,F_0\,|^2\sin^2(\omega t + \eta - \rho)/|\,Z\,|^2, \qquad (3.27)$$

where k has been written as $m\omega_0^2$ in the last expression. These expressions resemble those in Chapter 1 that represented T and U in the free motion of an undamped oscillator, in that both functions are positive definite, both oscillate with double the frequency of the motion, and their oscillations are 180° out of phase with each other. But in this case T and U do not represent the whole energy balance and flow; work is being done by the applied force and the force of friction. Correspondingly, T and U have different amplitudes of oscillation except at resonance, so in general their sum is time-dependent:

$$T + U = E = \tfrac{1}{4}\,m\,|\,\mathrm{F}_0\,|^2\,[(1 + \omega_0^2/\omega^2)$$
$$+ (1 - \omega_0^2/\omega^2)\cos(2\omega t + 2\eta - 2\rho)]/\,|\,Z\,|^2. \qquad (3.28)$$

It is readily verified that the time derivative of this expression for total stored energy E equals the rate at which work is being done on the moving mass, $vF - v^2R$, where the first term is the product of the real $v(t)$ and the real applied force, and the second is the (negative) product of $v(t)$ by the frictional force $-Rv(t)$.

At resonance E is constant in time, or at each instant the power input from F equals the power dissipation in friction: $vF = v^2R$, or $F = vR$. This result was anticipated in the fact that, at resonance, $Z = R$, whereas at all frequencies $Zv = F$.

All the foregoing arguments and results can easily be translated into electrical terminology by means of the correspondences displayed in Equations 1.22 and 1.46. The electrical circuit being described is shown in Figure 3.2. The electrical impedance of this circuit is

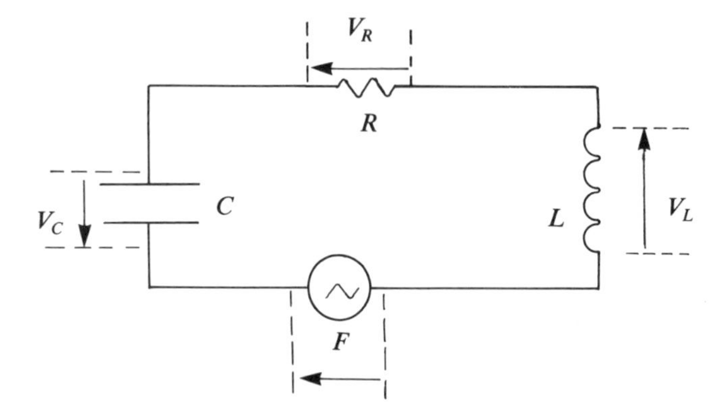

Figure 3.2. A sinusoidally driven RLC circuit.

$$Z = R + iX = R + i(\omega L - 1/\omega C) = Z_R + Z_L + Z_C. \tag{3.29}$$

The law of Kirchhoff that we have used already now states that

$$F = V_R + V_L + V_C, \tag{3.30}$$

i.e., that the applied EMF is the sum of the voltage drops around the circuit. In the case of sinusoidally varying steady-state current, I,

$$\begin{aligned} V_R &= IR = Z_R I \\ V_L &= i\omega\, LI = Z_L I \\ V_C &= I/i\omega\, C = Z_C I. \end{aligned} \tag{3.31}$$

Thus, since

$$F = ZI, \tag{3.32}$$

we can check the correctness of Equation 3.29, which gives Z for the series combination as the sum of the (complex) impedances of the separate circuit elements.

These relationships can be conveniently represented by means of a vector diagram, in which F, I, V_R, V_L, and V_C is each represented by a rotating vector in the complex plane (Figure 3.3). This diagram shows the situation at resonance ($\omega = \omega_0$), $V_L = -V_C$. At lower frequencies $|V_L| < |V_C|$ and at higher frequencies $|V_L| > |V_C|$.

With the help of Figures 3.4 and 3.5 we can draw some general conclusions about impedances of more complicated circuits. Figure 3.4 depicts two (unspecified) circuit elements, having complex impedances Z_1 and Z_2, in series. By the same sort of reasoning already used, we can set the total voltage difference across the series combination equal to the sum of the voltage differences across the two constituent parts:

$$ZI = V = V_1 + V_2 = (Z_1 + Z_2)I, \text{ or } Z = Z_1 + Z_2. \tag{3.33}$$

Note that these impedances are complex, being so defined as to give the correct relationship of amplitude and phase between real current and real voltages. It is correct to add the complex

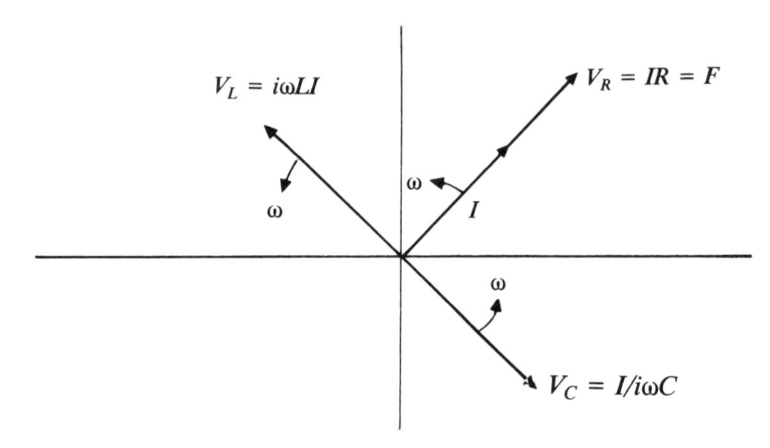

Figure 3.3. Amplitude and phase relationships of voltages and current in an RLC circuit driven at its resonant frequency.

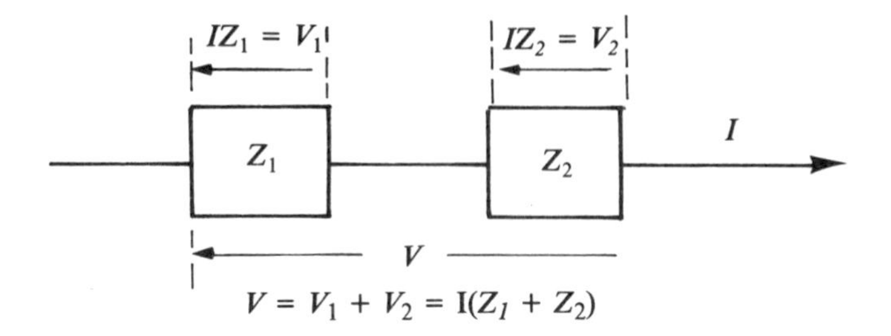

Figure 3.4. Two impedances connected in series.

voltages in Equation 3.33, because the real part of a sum is the sum of real parts.

Figure 3.5 shows two circuit elements in parallel. Instead of having a common current as in the previous case, they now have a common potential difference V, and separate currents I_1 and I_2 that add (again as complex functions) to produce a total current I. Now, defining *admittance*, Y, as $1/Z$, we have

$$YV = I = I_1 + I_2 = (Y_1 + Y_2)V,$$

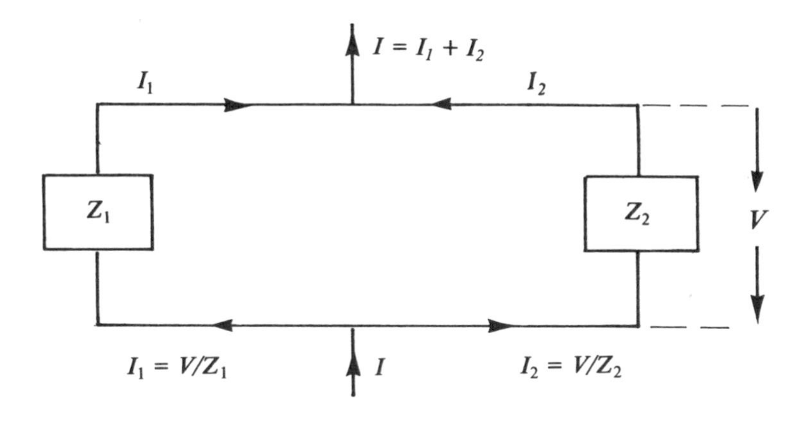

Figure 3.5. Two impedances connected in parallel.

or

$$Y = Y_1 + Y_2$$

or (3.34)

$$1/Z = 1/Z_1 + Z_2,$$

so

$$Z = Z_1Z_2/(Z_1 + Z_2),$$

which is the formula for combining impedances of circuit elements that are connected in parallel. The rules of combination in Equations 3.33 and 3.34 can be used to yield expressions for impedances (or admittances) of more complicated circuits.

The complex impedance between any two points in a circuit can always be written in terms of resistance and reactance:

$$Z = R + iX; \tag{3.35}$$

R and X, defined to be real, will in general have more complicated dependence on frequency then do the R and X of the simple series circuit discussed in this chapter. In all cases, however, the voltage between the two points and the total current flowing through the part of the circuit between the points (in at one point and out at the other) are related by

$$V = ZI, \tag{3.36}$$

where V, Z, and I are complex quantities. Now the rate of energy input to the part of the circuit between the two points is the product of the *real* V and the *real* I. The real part of V is a sum of two parts, one, in phase with the real part of I, coming from the product RI, and the other, coming from XI, 90° out of phase with the real I. When one multiplies the real I by this sum, one gets one term, containing R, which is proportional to the square of a sinusoidal function of time, whereas the X term is proportional to the product of two such functions that are 90° out of phase with each other. Only the R term will contribute to the time average of energy input, taken over a whole number of cycles; the X term will average to zero. Thus it is the *resistance* that causes energy dissipation, that is, net energy input to a part of a circuit, with the energy not being returned from that part to the rest of the circuit. The effect of *reactance* is to produce alternating storage and return of energy, whereby the reactive part of energy input averages to zero.

Sometimes it is desirable to choose a load impedance so as to maximize the power dissipated in it when it is connected in series with a fixed voltage generator at some frequency ω and having some fixed internal impedance Z_i. Let us then assume a voltage source F in series with an internal impedance $Z_i = R_i + iX_i$ and with a load impedance $Z = R + iX$, and let us take only R and X to be adjustable, F, R_i, X_i, and ω being fixed. Then Equation 3.36 gives the complex current flowing in the circuit, and, according to the subsequent argument, the time-average rate of energy dissipation in the load impedance is

$$\tfrac{1}{2}|I|^2R = \tfrac{1}{2}|F|^2R/|Z + Z_i|^2$$
$$= \tfrac{1}{2}|F|^2R/[(R + R_i)^2 + (X + X_i)^2]. \qquad (3.37)$$

Taking $|F|$, R_i, X_i fixed, we can maximize this quantity with respect to R and X. It is easy to show that maximum power dissipation in Z occurs when $R = R_i$ and $X = -X_i$, or

$$Z = Z_i^*. \qquad (3.38)$$

The process of choosing Z so as to maximize power dissipation in the external circuit (or power transfer from the generator to the load) is known as *impedance matching.* Although the example contemplated here is a simple one, the idea of impedance matching for maximum power transfer has quite broad applicability to linear systems of many types, including some in which power is carried by waves rather than current in wires.

NOTES

See references such as:

1. Kip, A. F. *Fundamentals of Electricity and Magnetism.* 2d ed. New York: McGraw Hill, 1969.
2. Pugh, E. M. and E. W. Pugh. *Principles of Electricity and Magnetism.* 2d ed. Reading, MA: Addison Wesley, 1970.
3. Symon, K. R. *Mechanics.* 3d ed. Reading, MA: Addison Wesley, 1971.

PROBLEMS

3.1 Verify explicitly that the choice of J given in Equation 3.16 causes the solution in Equation 3.9 to satisfy the indicated initial conditions.

3.2 Check the statements below Equation 3.28, about energy balance.

3.3 An overdamped series RLC circuit is driven by a sinusoidal voltage of frequency ω_0, and has reached its steady state of motion. Compare the peak values and phases of the potential differences across the three elements, R, L, and C.

3.4 A critically damped oscillator is subjected to the force

$$F(t) = A \text{ for } -\infty < t \leqslant 0.$$

$$F(t) = 0 \text{ for } t > 0.$$

Calculate the motion of the oscillator after $t = 0$.

3.5 A generator of frequency ω is driving the steady-state current through a circuit that contains R, L, and C in series. The internal resistance of the generator is r, which is thus effectively in series with the rest of the circuit. Taking ω, r, L, C, and the peak voltage of the generator as fixed, calculate the value of R that maximizes power dissipation in the part of the circuit outside of the generator.

3.6 A series RLC circuit is in its steady state of response under the influence of a sinusoidal applied emf. Then the emf is suddenly shorted out, i.e., the generator is replaced by a zero resistance. At what part of the cycle of the current should this shorting-out occur in order to maximize the energy dissipated in the damped motion that ensues? Does this optimal phase depend on the frequency of the steady-state current that precedes the shorting-out?

3.7 A hanging coil spring has a length of 10 cm with nothing hanging from it, and 20 cm with a 1-kg mass hanging on it. While this mass is hanging on the spring, the spring's point of suspension is moved sinusoidally up and down with amplitude 1 cm. Determine the amplitude and phase of the mass's steady-state motion if the frequency of motion of the point of suspension is 1 Hz, 10 Hz.

3.8 A series *RLC* circuit is driven by a current-generator, which forces a sinusoidal current of peak value I_0 and frequency ω to flow in the circuit. What are the time-dependent voltage differences across the three circuit elements?

4

Sums of Sinusoidal Forces or EMF's—Fourier Analysis

In Chapter 3 we studied the response on an oscillator to a driving force with sinusoidal time-dependence. In the present chapter we build up more complicated forces by adding sinusoidal forces together.

If an oscillator is driven by a sum of two forces like that discussed in the preceding chapter, the general solution of its equation of motion can be written

$$x(t) = x_H(t; A, B) + x_1(t) + x_2(t), \qquad (4.1)$$

where $x_1(t)$ is a particular solution of the equation with the force $F_1(t)$ on the right side, and $x_2(t)$ is such a solution with $F_2(t)$ as the driving force, and $x_H(t; A, B)$ is, as in the last chapter, the general solution of the homogeneous equation. In particular, if F_1 and F_2 are complex exponential functions like that used in Chapter 3, $x_1(t)$ and $x_2(t)$ are the corresponding steady-state solutions. The result can readily be extended to the case of more than two forces. The steady-state motion caused by a sum of sinusoidal forces is the sum of the steady-state motions caused by the separate forces.

One particular type of sum has been much studied, and much utilized in various applications, the Fourier series. This is a series (generally infinite) of sine and cosine terms, or complex exponential terms, all the terms having frequencies that are multiples of a highest common divisor known as the *fundamental* frequency. Three forms in which Fourier series are written are

$$f(t) = \sum_{n=0}^{\infty} a_n \cos n\Omega t + \sum_{n=1}^{\infty} b_n \sin n\Omega t;$$

$$f(t) = \sum_{n=0}^{\infty} c_n \cos (n\Omega t + \theta_n); \qquad (4.2)$$

$$f(t) = \sum_{n=-\infty}^{\infty} \alpha_n e^{in\Omega t}.$$

Using the fact that the sine and the cosine function of a given frequency are linearly independent, and that such functions of different frequencies are linearly independent also, one can easily work out the relationships among the quantities a_n, b_n, c_n, θ_n, α_n. One can verify, also, that real $f(t)$ implies real values of the a_n, b_n, c_n, θ_n, and implies that $\alpha_{-n} = \alpha_n^*$.

In each of the forms 4.2, the fundamental frequency is Ω, and the corresponding period is $\tau = 2\pi/\Omega$. Every term with $|n| > 1$ has a frequency that is a multiple of Ω, and an integral number of periods in the fundamental period τ. Therefore, if these series converge at all, they converge to a periodic function whose period is the least common multiple of the periods of the nonzero terms—usually τ. Therefore, Fourier series are used to represent functions that are periodic or that are defined only on a finite interval (so that the series' repetition on both sides of that interval does not matter).

We shall confine our attention to the third of the forms 4.2, the exponential form—though many of our remarks will apply, *mutatis mutandis*, to the first form. The exponential functions $e^{in\Omega t}$ have the convenient property of being mutually *orthogonal* in any interval of length τ. Our definition of the word orthogonal, sufficiently general for our purposes,[1] follows:

Two functions $u(t)$ and $v(t)$, defined over the interval $[a, b]$, are orthogonal in that interval if

$$(u, v) \equiv \int_b^a u^*(t)v(t)dt = 0. \tag{4.3}$$

Here (u, v), defined to be equal to the integral, is known as the "scalar product" of the functions u and v. According to the definition, this sort of multiplication is distributive with respect to addition, but is not commutative:

$$(u, v) = (v, u)^*. \tag{4.4}$$

This noncommutativity does not affect the definition of orthogonality; $(u, v) = 0$ implies $(v, u) = 0$.

If for u and v we use two of our exponential functions, $e^{in\Omega t}$ and $e^{im\Omega t}$, respectively, and for a and b we take any two times separated by the interval $\tau = 2\pi/\Omega$, we find that

$$\begin{aligned} (u, \mathrm{v}) &= \int_{t_0}^{t_0+\tau} e^{i(m-n)\Omega t}\, dt = 0 \text{ if } m \neq n \\ &= \tau = 2\pi/\Omega \text{ if } m = n. \end{aligned} \tag{4.5}$$

The integral vanishes when $m \neq n$ because $m - n$ is an integer and therefore the integration covers an integral number of periods of the cosine and sine functions that make up the exponential, which functions thus integrate to zero. The square root of the scalar product of a function with itself, as given on the second line of Equation 4.5, is known as the *norm* of the function, defined to be positive (or zero in the trivial case in which the function is zero almost everywhere in the interval).

We have called orthogonality a "convenient" property because of the following possibility; multiply the equation

$$f(t) = \sum_{n=-\infty}^{\infty} \alpha_n e^{in\Omega t} \tag{4.2'}$$

by the complex conjugate of one of the exponential functions, $e^{-ik\Omega t}$, and then integrate both sides over one fundamental period. Denoting $e^{in\Omega t}$ by $u_n(t)$, we find that

$$(u_k, f) = \int_{t_0}^{t_0+\tau} f(t)e^{-ik\Omega t}\,dt$$

$$= \int_{t_0}^{t_0+\tau} e^{-ik\Omega t}\left[\sum_{n=-\infty}^{\infty} \alpha_n e^{in\Omega t}\right] dt \tag{4.6}$$

$$= \sum_n \alpha_n \int_{t_0}^{t_0+\tau} e^{i(n-k)\Omega t}\,dt = \sum_n \alpha_n (u_k, u_n) = \alpha_k \tau.$$

Note that the integral of the sum is equated to a sum of integrals; the isolation of one coefficient α_k was possible only by use of this step. A sufficient condition for the correctness of such term-by-term integration of an infinite series is that the series converges uniformly in the interval of integration. In fact, $f(t)$ may have a finite number of discontinuities in the interval, provided it has bounded variation and its integral over the interval converges absolutely.[2] Use of the coefficients α_k determined by Equation 4.6 guarantees that the series converges to $f(t)$ at all points where f is continuous, and, at a discontinuity, to the mean of the values approached by f below and above the discontinuity. Thus, subject to these restrictions, any function $f(t)$ can be expressed as a Fourier series in any finite interval, and the coefficients in the series can be evaluated one by one because the functions being used are mutually orthogonal; the coefficients are

$$\alpha_k = (u_k, f)/(u_k, u_k) = \frac{1}{\tau}\int_{t_0}^{t_0+\tau} f(t)e^{-ik\Omega t}\,dt. \tag{4.7}$$

To clarify the foregoing results, we shall work out two examples. The first is

$$f(t) = 0 \qquad (\text{for } -\tau/2 < t \leqslant -t_0);$$

$$f(t) = A \qquad (\text{for } -t_0 < t < t_0); \tag{4.8}$$

$$f(t) = 0 \qquad (\text{for } t_0 \leqslant t \leqslant \tau/2);$$

$$f(t + \tau) = f(t) \text{ (for all } t).$$

Then

$$\alpha_n = \frac{1}{\tau}\int_{-\tau/2}^{\tau/2} f(t)e^{-in\Omega t}\,dt =$$

$$A\Omega(e^{-int_0\Omega} - e^{int_0\Omega})/(-in\Omega)2\pi$$

$$= (A\Omega\; t_0/\pi)\;(\sin n\Omega t_0)/n\Omega t_0. \qquad (4.9)$$

Thus

$$f(t) = (A\Omega t_0/\pi) \sum_{n=-\infty}^{\infty} e^{in\Omega t}\,(\sin n\Omega t_0)/n\Omega t_0$$

$$= (2A\Omega t_0/\pi) \sum_{n=0}^{\infty} (\cos n\Omega t\,\sin n\Omega t_0)/n\Omega t_0 \qquad (4.10)$$

$$= (A/\pi) \sum_{n=0}^{\infty} [\sin n\Omega(t_0 - t) + \sin n\Omega(t_0 + t)]/n.$$

Note that the coefficients are all real, so the series reduces to a series of cosines (middle form, Equation 4.10) as is appropriate for an even function $f(t)$. The last form is shown because it may sometimes be useful.

The second example is

$$f(t) = -Ae^{bt} \qquad (\text{for } -\tau/2 < t \leqslant 0);$$

$$f(t) = Ae^{-bt} \qquad (\text{for } 0 < t \leqslant \tau/2); \qquad (4.11)$$

$$f(t + \tau) = f(t) \text{ (for all } t).$$

Then it is straightforward to show that

$$\alpha_n = inA\Omega^2\,[(-1)^n\,e^{-b\tau/2} - 1]/(b^2 + n^2\Omega^2), \qquad (4.12)$$

and that

$$f(t) = 2A\Omega^2 \sum_{n=1}^{\infty} n \sin n\Omega t[1 - (-1)^n e^{-b\tau/2}]/(b^2 + n^2\Omega^2), \tag{4.13}$$

where in this case the imaginary coefficients produce a series of sine functions which appropriately converges to an odd function of t.

The term "orthogonal" means "perpendicular"; it is applied to functions by analogy with vector algebra in N-space. A set of orthogonal functions is analogous to a set of orthogonal vectors. If there are as many such vectors as the dimensionality of the space, no new vector can be found that is linearly independent of all of them, or in other words any vector in the space can be expressed in terms of the members of the set:

$$\mathbf{A} = \sum_{i=1}^{N} A_i \mathbf{v}_i, \tag{4.14}$$

where $\mathbf{A}$ is an arbitary vector, and the $\mathbf{v}_i$ are the members of the orthogonal set

$$\mathbf{v}_i \cdot \mathbf{v}_j = 0 \text{ if } i \neq j. \tag{4.15}$$

Then the components of $\mathbf{A}$, the A_i, can be isolated by means of scalar multiplication by the respective members of the orthogonal set

$$\begin{gathered} \mathbf{A} \cdot \mathbf{v}_j = \left(\sum_i A_i \mathbf{v}_i\right) \cdot \mathbf{v}_j = \sum_i A_i(\mathbf{v}_i \cdot \mathbf{v}_j) = A_j(\mathbf{v}_j \cdot \mathbf{v}_j); \\ A_j = (\mathbf{A} \cdot \mathbf{v}_j)/(\mathbf{v}_j \cdot \mathbf{v}_j). \end{gathered} \tag{4.16}$$

Thus the vector $\mathbf{A}$ is analogous to the function $f(t)$, the $\mathbf{v}_i$ are analogous to the orthogonal functions $u_i(t)$, and the dot product of vectors (for example, $\mathbf{v}_j \cdot \mathbf{A}$) is analogous to the scalar product of functions such as (u_j, f). In the case of finite dimensionality there is no problem of legitimacy in the use of the distributive law, which converts a dot product of a vector with a sum into a sum of dot products. When the dimensionality of a vector space goes to infinity, with scalar product being defined at each stage of the process, the distributive law applies only in suitable circumstances which are analogous to the conditions for term-by-term integration of an infinite series. What we have been calling an "analogy" between vectors and functions is really an isomorphism, and one speaks of function spaces, and of a set of functions "spanning" such a space (which means that the set is *complete*, that there are no additional functions in the space that are linearly independent of those already part of the set).

Having defined Fourier series and discussed how to solve for the coefficients, we can now apply them to the problem of the oscillator driven by a periodic force. The equation of motion is

$$\ddot{x} + 2\gamma\dot{x} + \omega_0^2 x = F(t)/m = f(t), \tag{4.17}$$

and we assume that, from $t = -\infty$ to $t = \infty$, $f(t)$ is the periodic function

$$f(t) = \sum_{n=-\infty}^{\infty} \alpha_n e^{in\Omega t}, \tag{4.18}$$

with the coefficients being given by Equation 4.7. Now the motion $x(t)$ is simply the steady-state motion which, as indicated at the beginning of this chapter, is the sum of steady-state motions due to the separate exponential terms in $f(t)$. We can determine this motion easily by expressing $x(t)$ as a Fourier series and substituting it into Equation 4.17:

$$\begin{aligned} x(t) &= \sum_{n=-\infty}^{\infty} \beta_n e^{in\Omega t}, \\ \dot{x}(t) &= \sum_{n=-\infty}^{\infty} \beta_n (in\Omega) e^{in\Omega t}, \\ \ddot{x}(t) &= \sum_{n=-\infty}^{\infty} \beta_n (-n^2\Omega^2) e^{in\Omega t}, \end{aligned} \tag{4.19}$$

where we have differentiated the series term by term—a procedure that is legitimate provided the highest derivative thus obtained satisfies the conditions for term-by-term integration. Now, moving $f(t)$ to the left side of Equation 4.17 and substituting Fourier series in all the terms, we have

$$\sum_{n=-\infty}^{\infty} [(-n^2\Omega^2 + 2in\gamma\Omega + \omega_0^2)\beta_n - \alpha_n]e^{in\Omega t} = 0. \qquad (4.20)$$

Thus we have a Fourier series representing the function 0; by Equation 4.7, with $f(t) = 0$, each coefficient in the series separately equals zero:

$$\beta_n = \frac{1}{-n^2\Omega^2 + 2in\Omega\gamma + \omega_0^2}\,\alpha_n, \qquad (4.21)$$

so

$$x(t) = \sum_{n=-\infty}^{\infty} \frac{\alpha_n e^{in\Omega t}}{-n^2\Omega^2 + 2in\Omega\gamma + \omega_0^2}. \qquad (4.22)$$

Here, just by assuming for $x(t)$ the Fourier series form, with fundamental frequency Ω and fundamental period $\tau = 2\pi/\Omega$, we have determined $x(t)$ uniquely. The constants of integration were in the transient term x_H which has long since died out, and the surviving solution is, as expected, the sum of steady-state solutions caused by the separate terms in the series for $f(t)$. Note that the series for x converges more strongly than that for $\dot{x}$, which in turn converges more strongly than that for $\ddot{x}$ (Equation 4.19), whereas the $\ddot{x}$ series, according to Equations 4.19 and 4.22, converges at the same rate as the $f(t)$ series. Thus, if it was legitimate to integrate the series for $f(t)$ term by term, it was legitimate also to differentiate the $x(t)$ series twice in the same manner.

Although Equation 4.22 gives the response of the oscillator to any periodic driving force, it is seldom possible to sum such a series in closed form. If obtaining numerical solutions of particular problems is required, it is usual to evaluate terms one by one, starting with $n = 0$, then including $n = 1, -1$, then $n = 2, -2$, and so on, carrying out the process until the terms are small enough to indicate that the partial sum is accurate enough. Such a process can go very quickly if done by a computer. Note that, if $f(t)$ is real, so that $\alpha_n = \alpha_{-n}$, it follows that $\beta_n = \beta_{-n}^*$, whence $x(t)$ and its derivatives are real also; it is not necessary to take real parts, for the imaginary parts add up to zero. Indeed, the sum of the nth and the $(-n)$th terms is real, for each n separately.

Fourier series can be used to represent, not only periodic functions but also functions that are defined only in finite intervals. For physical reasons, functions of time are seldom of this type, so in the problems that we are studying Fourier series are applicable only when driving forces or emfs are periodic functions. There are many interesting driving forces that are nonperiodic, so we shall next extend our treatment to include such cases.

The strategy that we shall use is to let the fundamental period of the Fourier series increase without limit, so that it covers the whole t-axis from $-\infty$ to $+\infty$. Then the series (now an integral) represents a function with infinite period, or a nonperiodic function, which is defined over the whole t-axis. To this end we shall choose the arbitrary time t_0 in Equation 4.7 to be $-\tau/2$, so the interval is symmetric about $t = 0$. We also substitute from Equation 4.7 in Equation 4.18:

$$f(t) = \sum_{n=-\infty}^{\infty} \frac{\Omega}{2\pi} e^{in\Omega t} \int_{-\tau/2}^{\tau/2} f(t') e^{in\Omega t'}\, dt', \tag{4.23}$$

where the coefficient α_n is now explicitly written as a time integral (in which the variable of integration is distinguished from t by means of a prime). Now, letting

$$\omega = n\Omega, \tag{4.24}$$

we note that the change of ω from one term of the series to the next, as n increases by 1, is

$$\Delta\omega = \Omega, \tag{4.25}$$

which we write in the series explicitly. Now as we let Ω approach zero and ω approach infinity, the sum over ω values spaced by the amount Ω goes over into a definite integral:

$$f(t) = \frac{1}{2\pi}\int_{-\tau/2}^{\tau/2} e^{i\omega t}\,\Delta\omega\left[\int_{-\tau/2}^{\tau/2} f(t')e^{-i\omega t'}\,dt'\right]$$

$$\underset{\Omega\to 0}{\longrightarrow}\frac{1}{2\pi}\int_{-\infty}^{\infty} e^{i\omega t}\,d\omega\left[\int_{-\infty}^{\infty} f(t')e^{-i\omega t'}\,dt'\right]. \tag{4.26}$$

This identity, the limiting form of the identity 4.23, is satisfied by any function f of a large class. It can be construed as a relationship between two functions f and α, like the relationships 4.7 and 4.18 between a function f and a set of coefficients α_n. Now the coefficients have become uncountable, corresponding to the points on the continuous ω-axis, and (except for a factor $[2\pi]^{-1/2}$) are collectively called $\alpha(\omega)$:

$$f(t) = \frac{1}{\sqrt{2\pi}}\int_{-\infty}^{\infty} \alpha(\omega)e^{i\omega t}\,d\omega;$$

$$\alpha(\omega) = \frac{1}{\sqrt{2\pi}}\int_{-\infty}^{\infty} f(t)e^{-i\omega t}\,dt. \tag{4.27}$$

These integrals are known as *Fourier integrals;* jointly they define the *Fourier transformation*. The functions f and α, related by the Fourier transformation, are known as *Fourier transforms* of each other. Their relationship is strikingly symmetric, unlike that of the periodic $f(t)$ and its Fourier coefficients α_n; here the only failure of symmetry between f and α is in the signs of the respective exponents in the integrals.

We shall now work out a few examples of Fourier integrals, both for the sake of illustrating the procedure and for the interest and usefulness of the results. First let us look at

$$f(t) = 0 \text{ for } t < 0,$$
$$f(t) = Ae^{-bt} \text{ for } t \geqslant 0. \tag{4.28}$$

Then

$$\alpha(\omega) = (2\pi)^{-1/2} \int_{-\infty}^{\infty} f(t)e^{-i\omega t}\, dt = (2\pi)^{-1/2} \int_{0}^{\infty} Ae^{-(b+i\omega)t}\, dt$$
$$= A/\sqrt{2\pi}\,(b + i\omega) = -Ai/\sqrt{2\pi}\,(\omega - ib). \tag{4.29}$$

Note that, although $f(t)$ is defined by different expressions on different parts of the t-axis and is discontinuous at $t = 0$, it has a single Fourier transform (independent of t) which happens to be a continuous function of ω. If this $\alpha(\omega)$ is substituted into the upper integral in 4.27, it gives back the $f(t)$ defined in 4.28—a fact that we shall later prove, though it may at present appear incredible.

As our next example let us take

$$f(t) = 0 \text{ for } t < -t_0 \text{ and for } t > t_0,$$
$$f(t) = A \text{ for } -t_0 \leqslant t \leqslant t_0. \tag{4.30}$$

The transformation is straightforward, and yields

$$\alpha(\omega) = (2/\pi)^{1/2}\, t_0 A \,\sin\omega\, t_0/\omega t_0, \tag{4.31}$$

which is again a single continuous function of ω, independent of t.

The third example, a Gaussian function, requires a short digression on the integration of such functions. Let I_n be as defined below, n being a nonnegative integer, and change variables to $y = x^2$:

$$I_n = \int_0^{\infty} x^n e^{-x^2}\, dx = \tfrac{1}{2}\int_0^{\infty} e^{-y}\, y^{(n-1)/2}\, dy$$
$$= \tfrac{1}{2}\Gamma\left(\frac{n+1}{2}\right) = \tfrac{1}{2}\left(\frac{n-1}{2}\right)! \text{ (if } n \text{ is odd).} \tag{4.32}$$

Here the symbol Γ denotes the gamma function, which is usually defined as an integral like the one above, and which is a generalization of the factorial:

$$\Gamma(k) = (k - 1)! \text{ if } k \text{ is a positive integer.} \tag{4.33}$$

It is easy to prove by partial integration that, for all k for which the integrals converge,

$$\Gamma(k + 1) = k\Gamma(k). \tag{4.34}$$

This relationship, plus the easily verified fact that $\Gamma(1) = 1$, suffices to determine the Γ-function for all positive integers and to establish its relationship to the factorial. Thus we have worked out the value of I_n for odd values of n.

To work it out for even n, we have to determine the gamma function for some k which is an integer plus 1/2. There is a convenient trick whereby we can get I_0, which is $(1/2)\,\Gamma(1/2)$:

$$I_0 = \int_0^\infty e^{-x2}\,dx, \text{ so } I_0{}^2 = \left[\int_0^\infty e^{-x2}\,dx\right]\left[\int_0^\infty e^{-y2}\,dy\right]$$

$$= \int_0^\infty \int_0^\infty e^{-(x^2+y^2)}\,dx\,dy \tag{4.35}$$

$$= \int_0^\infty dr \int_0^{\pi/2} r\,d\theta\; e^{-r^2},$$

where the last form results from a change of variables in the xy-plane from (x, y) to polar coordinates (r, θ). The θ integration can be done immediately, yielding a factor $\pi/2$, and the r integration goes easily with the aid of the substitution $s = r^2$. The result is

$$I_0^2 = \tfrac{1}{4}\pi, \text{ so } I_0 = \tfrac{1}{2}\Gamma(\tfrac{1}{2}) = \tfrac{1}{2}\sqrt{\pi}, \tag{4.36}$$

whence one can work out the other even I_n.

To evaluate the integral of $x^n e^{-x^2}$ from $-\infty$ to ∞, note that even n implies an even function, whereby the desired integral is $2I_n$; odd n yields zero. Note, too, that

$$\int_0^\infty x^n e^{-(ax)^2} dx = a^{-n-1} I_n. \tag{4.37}$$

Now let us generalize the case with $n = 0$ by including a linear term in the exponent; let us evaluate

$$J = \int_{-\infty}^{\infty} e^{-a^2x^2-bx} dx. \tag{4.38}$$

This can be done by completion of the square; let $y = ax + c$, and choose c so that $y^2 = a^2x^2 + bx + c^2$. This means that $2ca = b$; also $dx = dy/a$. Then

$$J = (e^{c^2}/a) \int_{-\infty}^{\infty} e^{-y^2} dy = 2I_0 e^{(b/2a)^2}/a = \frac{\sqrt{\pi}}{a} e^{(b/2a)^2}. \tag{4.39}$$

This result can be applied directly to the Fourier transformation of a Gaussian function. Let

$$f(t) = Ae^{-a^2t^2}. \tag{4.40}$$

This its transform is

$$\alpha(\omega) = \frac{A}{\sqrt{2\pi}} \int_{-\infty}^{\infty} e^{-a^2t-i\omega t} dt = \frac{A}{\sqrt{2}a} e^{-\omega^2/4a^2}, \tag{4.41}$$

a function of ω having the same form as $f(t)$, but with a playing an opposite role.

All our examples of Fourier series and integrals can be appreciated better with the help of the idea of a *frequency spectrum.* In the Fourier series a function is being represented as a sum of many functions with different frequencies $\omega_n = n\Omega$; in the integral the frequencies are a continuous rather than a discrete set. In each case there is a complex function of frequency—α_n in the series, $\alpha(\omega)$ in the integral—which expresses the amplitude and the phase of the component of $f(t)$ at each of the frequencies that enter f. This function of frequency is called the *frequency spectrum* of the function $f(t)$—a discrete spectrum in the case of a periodic $f(t)$, a continuous spectrum if $f(t)$ is nonperiodic. A spectrum of either type must approach zero rapidly enough as $\omega \to \pm\infty$ to ensure convergence of the series or integral for $f(t)$. The rapidity with which the spectrum approaches zero is determined by the smoothness of $f(t)$; an $f(t)$ without cusps or discontinuities can be fitted with smaller amounts of the high-frequency functions, whereas such functions are needed in larger amounts if the $f(t)$ being fitted has sharp corners. The two examples of Fourier series and the first two examples of Fourier integrals all have discontinuities; correspondingly, their spectra go very slowly to zero at $\omega = \pm\infty$, like $1/\omega$. On the other hand, the third example of the integral, the Gaussian, has an $f(t)$ without discontinuities and a corresponding $\alpha(\omega)$ that goes very rapidly to zero at $\pm\infty$. Furthermore, the width of the spectrum is inversely proportional to the width of $f(t)$. The $1/e$ points of $f(t)$ (where $f(t)$ equals its maximum value over e) are at $t = \pm 1/a$, so large a implies a narrow peak of $f(t)$; the $1/e$ points of $\alpha(\omega)$ are at $\omega = \pm 2a$, so large a implies a broad spectrum, $\alpha(\omega)$, of $f(t)$. To fit a narrow peak of f one needs more high-frequency contributions than are necessary to fit a wide peak.

Although the Gaussian function is the only one of our examples in which the original function and its spectrum have the same mathematical form, one can see the same inverse relationship between the spreads of $f(t)$ and of $\alpha(\omega)$ also in the (rather similar) examples of the series and the integral, defined in Equations 4.8 and 4.30. Each of these cases involves a "square" pulse of duration $2t_0$ and amplitude A, periodic in one case and nonrepeating in the other. Let us now, in each case, set $A = 1/2t_0$, so the time integral of the pulse equals 1. Then

$$\alpha_n = (\Omega/2\pi)(\sin \Omega t_0)/n\Omega t_0,$$

and (4. 42)

$$\alpha(\omega) = (2\pi)^{-1/2} \sin\omega t_0/\omega t_0,$$

which are exactly alike except for the factors in front, $n\Omega$ being the frequency Ω in the case of the series. Now clearly if we reduce t_0, making $f(t)$ in a sense "sharper," we find that the spectrum spreads out on the ω-axis. Going to the limit $t_0 = 0$ we encounter an anomalous situation; the spectrum spreads out so far as to have become independent of frequency entirely. Thus

$$\alpha_n \underset{t_0 \to 0}{\to} \Omega/2\pi;$$
$$\alpha(\omega) \underset{t_0 \to 0}{\to} (2\pi)^{-1/2}. \tag{4.43}$$

We have written these expressions as limits rather than as equalities for two related reasons: these constant "spectra" produce divergent series and integral, so they are not legitimate spectra but rather limits of sequences of legitimate spectra; and the limiting "functions" $f(t)$ are presumably infinitely narrow, infinitely high spikes with unit area—and there are no such functions. Thus, it would be wrong to say that the limiting (constant) "spectra" are the spectra of the limiting instantaneous functions. The constant "spectra" are not spectra of anything, for they do not produce convergence; and the limiting "functions" are not functions, for contradictory properties are attributed to them.

Despite these logical obstacles, physicists call the nonfunction a *delta function* and attribute to it a constant Fourier transform equal to $(2\pi)^{-1/2}$:

$$\delta(\tau) = (2\pi)^{-1/2} \int_{-\infty}^{\infty} (2\pi)^{-1/2} e^{i\omega t}\, d\omega;$$
$$(2\pi)^{-1/2} = (2\pi)^{-1/2} \int_{-\infty}^{\infty} \delta(t)\, e^{-i\omega t}\, dt. \tag{4.44}$$

The first of these equations makes sense of a sort, in that when $t \neq 0$ the integrand oscillates, moving around a circle in the complex ω-plane, so its average value is zero and the integral might be taken to be zero—but really the integral diverges. When $t = 0$, the integrand $= 1$ everywhere on the ω-axis, so the integral equals infinity—but really it does not equal anything. It might be considered appropriate for a nonexistent integral to be equated to a nonexistent function.

The second equation exemplifies what is often taken to be the defining property of the delta function:

$$\int_{<t'}^{>t'} \delta(t - t')g(t)dt = g(t'), \tag{4.45}$$

where g is any function defined at $t = t'$, and the integration over t includes the point t'. There is no function $\delta(t - t')$ that satisfies Equation 4.45 for all g, but functions like that defined in Equation 4.30 approximately satisfy it, and the approximation can be made as accurate as one wishes, short of perfection. Let

$$\begin{aligned} \delta_a(t - t') &= \tfrac{1}{2}a \text{ when } t' - a < t < t' + a \\ &= 0 \text{ elsewhere.} \end{aligned} \tag{4.46}$$

Then

$$\int_{<t'-a}^{>t'+a} \delta_a(t - t')g(t)dt = \bar{g}_a(t'), \tag{4.47}$$

which means the average of $g(t)$ in a region of width $2a$ centered at $t = t'$. The limit of this integral, then, as $a \to 0$, is $g(t')$, so Equation 4.45 is often written with the claim that it is the limit of Equation 4.47. But the integrand in Equation 4.47 does *not* approach any limit that can be identified with the "integrand" in Equation 4.45. The limit of the *integral* in Equation 4.47 has the value claimed for the symbol in Equation 4.45 that looks like an integral. Thus this symbol should not be interpreted as an integral, but as the limit of a sequence of integrals. In general, the symbol representing the delta "function" makes sense only afer an integral sign (or in an equation that is later going to be integrated)—and then the "integral" is really the limit of a sequence of integrals and not an integral itself.

Provided one understands these logical points, one can usefully employ a widely used shorthand, writing relationships like Equation 4.44 and 4.45 as if they meant what they appear to mean and as if there were a delta function. We shall do so hereinafter, taking care with the order of limits and integrals only when care is needed to prevent error.[3]

The example shown in Equations 4.28 and 4.29 has a limiting case that gives rise to some, but not all, of the difficulties associated with the case just discussed. When $b = 0$ in Equation 4.28, the function becomes $A\ \theta\ (t)$, where

$$\theta(t) = 0 \qquad \text{for } t < 0$$
$$= 1 \qquad \text{for } t > 0, \tag{4.48}$$

a *step function*. Unlike the delta function, the step function exists on the whole t-axis. Its Fourier transform approaches the limit $-Ai/\omega\sqrt{2\pi}$, again a well-defined function, but *not* the transform of $A\theta(t)$—which does not have a Fourier transform, because the integral diverges that should give the transform. Thus, when working with step functions, we should keep in mind the possibility that we may have to use care with limits. Incidentally, in the spirit in which we use delta functions, we can write, also,

$$d\theta(t)/dt = \delta(t). \tag{4.49}$$

With the foregoing examples, explanations, and cautions we have digressed somewhat from our main line of argument, which is intended to lead to use of the Fourier integral in the solution of mechanical and electrical problems. Let us now consider a damped oscillator acted on by a force that is expressible as a Fourier integral. If $f(t)$ = force/mass, we write

$$f(t) = (2\pi)^{-1/2} \int_{-\infty}^{\infty} \alpha(\omega)\, e^{i\omega t}\, d\omega. \tag{4.50}$$

As in the case of Fourier series, let us now also assume that the response of the oscillator can be expressed as a Fourier integral:

$$x(t) = \frac{1}{\sqrt{2\pi}} \int_{-\infty}^{\infty} \beta(\omega)\, e^{i\omega t}\, d\omega. \tag{4.51}$$

As we did in the case of the series, we now move the $f(t)$ term in Equation 4.17 to the left side, and substitute into the equation the expression above. Again we interchange the order of differentiation and summation or integration, a step that is justified by arguments like those following Equation 4.22. Then, combining the several Fourier integrals into one, we have a Fourier integral equal to zero; thus the integrand must equal zero, and it follows that

$$\beta(\omega) = \frac{\alpha(\omega)}{-\omega^2 + 2i\gamma\omega + \omega_0^{\,2}}, \tag{4.52}$$

and hence that

$$x\,(t) = \frac{1}{\sqrt{2\pi}} \int_{-\infty}^{\infty} \frac{-\alpha(\omega)e^{i\omega t}d\omega}{\omega^2 - 2i\gamma\omega - \omega_0^{\,2}}. \tag{4.53}$$

Note that, as in the case of the Fourier series, the assumption that $x(t)$ can be written in the form 4.51, as a Fourier integral, determines it uniquely. The constants of integration have been determined already. The motion has been calculated as a sum of steady-state responses to the various forces $\alpha(\omega)e^{i\omega t}\,d\omega$, each present from $t = -\infty$, that together make up the actual $f(t)$, so no transient motion $x(t)$ has been included explicitly. But, our $f(t)$ is a "temporary" sort of function, in that its expressibility as a Fourier integral requires it to go to zero at infinity. Thus, we expect that the entire response to $f(t)$ will be of the transient type; as $t \to \infty$, $x(t)$, like $f(t)$, must go to zero—otherwise, it cannot even be written in the form 4.51.

It seems reasonable to wonder whether a solution like $x(t)$, from which transient responses have been deliberately excluded, can possibly behave in the way that the response to a force like $f(t)$ has to behave. In order to answer this question, we should explicitly calculate some particular repsonses $x(t)$. For any $\alpha(\omega)$ that can conceivably enter it, however, the integral 4.53 is quite forbidding. We cannot, in general, evaluate indefinite integrals containing such integrands. Fortunately, it happens that many of the Fourier integrals arising in problems like the present one can be evaluated, by a process of integration in the complex ω-plane. We shall discuss this process in the next chapter.

In concluding this chapter, we shall briefly discuss a property of Fourier integrals that underlies many applications, including one application that we shall make in a later chapter. Let two functions of t be given by

$$f(t) = (2\pi)^{-1/2} \int_{-\infty}^{\infty} F(u)\, e^{iut}\, du, \tag{4.54}$$

$$g(t) = (2\pi)^{-1/2} \int_{-\infty}^{\infty} G(v)\, e^{ivt}\, dv,$$

Now let $h(t)$ be the product of f and g, and let us ask for the Fourier transform of h:

$$h(t) \equiv f(t)\, g(t) = (2\pi)^{-1/2} \int_{-\infty}^{\infty} H(w)\, e^{iwt} dw, \qquad (4.55)$$

and let us seek to identify H by multiplying the two Fourier integrals for f and g:

$$h(t) = f(t)\, g(t) = (2\pi)^{-1} \int_{-\infty}^{\infty} \int_{-\infty}^{\infty} F(u)\, G(v)\, e^{i(u+v)t}\, du\, dv$$

$$(4.56)$$

$$= (2\pi)^{-1} \int_{-\infty}^{\infty} e^{iwt} \left[\int_{-\infty}^{\infty} F(u)\, G(w-u)\, du \right] dw,$$

where in the last step we have changed variables in the double integral from (u, v) to (u, w), where $w = u + v$, and the Jacobian of the transformation equals one. Now, comparing 4.55 and 4.56, we can identify the transform of $h(t)$:

$$H(w) = (2\pi)^{-1/2} \int_{-\infty}^{\infty} F(u)\, G(w-u) du \equiv (2\pi)^{-1/2}\, F{*}G\, (w). \qquad (4.57)$$

This integral, denoted by $F{*}G$, is known as the *convolution* (German, *Faltung*) of the functions F and G. Thus we have shown that the Fourier transform of the product of two functions is the convolution of the separate transforms, times $(2\pi)^{-1/2}$. Note that this result is unchanged if applied to inverse transforms, in which the exponents have negative signs. Note also that the convolution operation is commutative: $F{*}G = G{*}F$. One speaks sometimes of the convolution operation as a "folding" (which is what "Faltung" means); the functions F and G are said to be "folded together." There are practical situations in which one of the functions is known, and $F*G$ is known also; and the problem consists of discovering what function has been folded in with the known function, i.e., of "unfolding" the known convolution. One way of solving this problem is to calculate the Fourier transforms of the known functions, divide one by the other, and then transform back.

We shall encounter convolutions later, when we discuss Green's functions.

NOTES

1. Some authors generalize the definition of orthogonality by including in the integrand of Equation 4.3 an additional factor $w(t)$, known as a "metric" or a "weighting" function. See, for example, Arfken, G. *Mathematical Methods for Physicists*. 2d ed. New York: Academic Press, 1970. Morse, P. M. and H. Feshbach. *Methods of Theoretical Physics*. New York; McGraw Hill, 1953.
2. We state many mathematical results without proof. Proofs can be found in books on mathematical analysis, such as E. T., Whittaker, and G. N. Watson. *A Course of Modern Analysis*. American ed, New York; Macmillan, 1943.
3. Mathematical analysis has been extended to include nonfunctions such as the delta "function," which is called a "distribution" instead of a function. The branch of analysis that deals with such objects is known as the theory of distributions. It is discussed in Barros-Neto, J. *An Introduction to the Theory of Distributions". New York; Dekker,* 1973. Challifour, J. L. *Generalized Functions and Fourier Analysis; an Introduction.* Reading, MA; Benjamin, 1972.
4. Useful general references on Fourier analysis are: Churchill R. V. and J. W. Brown. *Fourier Series and Boundary Value Problems*, 3d ed. New York: McGraw Hill, 1978. Campbell, G. A. and R. M. Foster. *Fourier Integrals for Practical Applications*. New York; Van Nostrand, 1947.

PROBLEMS

4.1. The periodic function $f(t)$ has period T. In one of its periods it is defined as

$$f(t) = At(t + T/2) \qquad (\text{for } -T/2 < t \leqslant 0)$$

$$f(t) = -At(t - T/2) \qquad (\text{for } 0 < t \leqslant T/2).$$

Express $f(t)$ as a Fourier series of exponential form, and then rewrite it in trigonometric form.

4.2. Starting with the $f(t)$ defined in Equation 4.11, replace it with $g(t)$ which is zero in the part of the period with $t < 0$ (line 1 of the definition of f), but equals $f(t)$ in the rest of that period, and has the same period as f. Express $g(t)$ in Fourier series, and compare the coeffiicents with the Fourier transform of the $f(t)$ defined in Equation 4.28.

4.3. Express in Fourier series $f(t) = A|\sin kt|$. What is the fundamental frequency?

4.4. Express in Fourier series:

$$f(t) = At \qquad (\text{for } -T/2 < t \leqslant T/2)$$

$$f(t + T) = f(t) \text{ (for all } t).$$

4.5. Express in Fourier series:

$$f(t) = A\cos(\omega_0 + E)t \qquad \text{for } -\frac{\pi}{\omega_0} < t \leqslant \frac{\pi}{\omega_0},\ |E| < \frac{\omega_0}{4}$$

$$f\left(t + \frac{2\pi}{\omega_0}\right) = f(t) \text{ (for all } t).$$

4.6. Express as Fourier integrals:

(a) $Ate^{-a^2t^2}$

(b) $A[e^{-a^2(t-t_0)^2} + e^{-a^2(t+t_0)^2}]$

(c) $A\cos^2 kt$ for $-n\pi/2k < t < n\pi/2k$, 0 elsewhere

(d) $A[\delta(t - t_0) + \delta(t + t_0)]$

(e) $\frac{1}{2}be^{bt}$ (for $t < 0$),
$\frac{1}{2}be^{-bt}$ (for $t > 0$).

4.7. Show that, by evaluating a "reverse" Fourier integral, one can recover Equation 4.40 from Equation 4.41.

4.8. Let

$$f(t) = \sum_{n=-\infty}^{\infty} \delta(t - 2\pi n).$$

Express $f(t)$ in Fourier series and Fourier integral, and check the two forms for equivalence.

4.9. Express as Fourier series

$$f(t) = A \sum_{n=-\infty}^{\infty} e^{-\alpha(t - 2\pi n)^2}$$

4.10. An underdamped oscillator is driven by the force

$$F(t) = A \text{ (for } -\tau/2 < t < 0);$$

$$F(t) = -A \text{ (for } 0 < t < \tau/2);$$

$$F(t + \tau) = F(t).$$

This force starts at very low fundamental frequency (large τ), and τ is slowly decreased with A constant, the variation being slow enough so the oscillator is at all times in its steady state of motion. At what values of τ (or Ω) will the oscillator's motion be particularly violent? Why? Will this strong motion always be at approximately the same frequency?

4.11. A pendulum of length S is swinging in a plane, with amplitude $A \ll S$. It is at $x = A$ when a wall is thrust into its path at $x = b$. The wall is perpendicular to the pendulum's motion; the pendulum rebounds from it with no loss of energy. Taking $x(0) = A$, express as a Fourier series the pendulum's $x(t)$ as affected by the wall.

4.12. A set of functions $u_n(at)$ is defined by the equation

$$e^{-s^2+2ast-a^2t^2/2} = \sum_{n=-\infty}^{\infty} \frac{s^n}{n!} u_n\,(at),$$

where t and s are both real variables and a is a real constant. Calculate the Fourier transform of $u_n(at)$, and show that it is proportional to the same function u_n of a multiple of ω. The functions u_n are proportional to the so-called Hermite functions.

5

Integration in the Complex Plane

In order to evaluate the intractable integrals arising from Equation 4.53, we need to devote some space first to the theory of functions of a complex variable. We shall present with no proof, or very little, many results that deserve careful study and rigorous treatment. The reader is advised to read one of the good books on the subject,[1] or to attend lectures on the subject by a mathematician.

By a complex variable we mean a variable, z, that can assume any complex value. We shall write z alternatively in real-imaginary form and in polar form:

$$z = x + iy = re^{i\theta};\ x, y, r, \theta \text{ real.} \tag{5.1}$$

By a function of z we mean another complex variable, w, whose real and imaginary parts (or whose amplitude and phase) are determined by the values of x and y:

$$w(z) = u(x, y) + iv(x, y);\ u, v \text{ real.} \tag{5.2}$$

Thus, in the sense that a complex number is an ordered pair of real numbers, a function of a complex variable is an ordered pair of functions of two real variables.

This definition of a function of a complex variable is too general to be interesting. By studying a restricted class of such functions, we shall arrive at many remarkable, useful results. Let us first inquire into the differentiation of $w(z)$ *with respect to z*. We shall define the derivative of w with respect to z precisely as the derivative of a function of a real variable is defined:

$$dw/dz = w'(z) = \lim_{h\to 0} [w(z + h) - w(z)]/h, \quad (5.3)$$

where h is a complex number, a member of a sequence that has the limit zero. Note that we are treating w as a function of a single variable, z, which has complex values, and we have defined the derivative accordingly. With the help of this derivative, we can write the partial derivatives of w with respect to x and y (denoted with subscripts) in different ways, according as u and v are treated as intermediate variables, or z is treated as a single intermediate variable:

$$\begin{aligned} w_x &= u_x + iv_x = w'z_x = w', \\ w_y &= u_y + iv_y = w'z_y = iw'. \end{aligned} \quad (5.4)$$

We can now equate the two expressions for $w' = dw/dz$ contained in these equations:

$$w_y = iw_x$$

or (5.5)

$$u_y + iv_y = -v_x + iu_x.$$

Now, u, v, x, y all being real by definition, we can separately equate real and imaginary parts in the last equation:

$$u_y = -v_x; \quad (5.6)$$

$$v_y = u_x.$$

These equations are known as the Cauchy-Riemann (C-R) equations. It is important to realize that they follow from the existence of the derivative $w'(z)$, and that they are not identities. If one chooses u and v at random, at differentiable functions of x and y, the chances are very great that the Equations 5.6 will *not* be satisfied. If they are not, we are forced to the conclusion that $w'(z)$ does not exist—at least, not at the points z where the equations fail. So it appears that a function $w(z)$ is unlikely to have a derivative with respect to z at any point, even though its real and imaginary parts may have an infinite number of successive derivatives with respect to x and y. Not only must u and v be differentiable, but also the very restrictive C-R conditions must be satisfied by their derivatives.

A function of a real variable can fail to be differentiable at a given point because the sort of ratio appearing in Equation 5.3 can fail to have a limit, or may have different limits when h approaches zero from different directions (above and below). When h is complex, it can approach zero in many more different ways than can a real h; thus a non-unique limit would appear more likely. If one examines the use of the chain rule involving w' in Equation 5.4, we can see that in one case w' refers to real values of h (increment of z parallel to the real axis), whereas in the other it involves imaginary values. Thus the use of the same w' in these two cases implies that the ratio in Equation 5.3 approaches the same limit regardless of whether h approaches zero through a sequence of real or of imaginary values. Therefore the C-R equations can be construed as guaranteeing that these two limits are the same.

In fact, it is easy to show that the C-R equations guarantee the uniqueness of the derivative also for other directions from which h may approach zero. If we change independent variables from z to $\tilde{z} = z\, e^{i\phi}$, where ϕ is a constant angle of rotation, we have either rotated every vector in the z-plane through an angle ϕ or we have left the vectors unchanged and rotated the x and y axes through an angle $-\phi$. Let us construe these transformation equations in the second (so-called "passive") way:

$$x = \tilde{x}\cos\phi + \tilde{y}\sin\phi$$

$$y = -\tilde{x}\sin\phi + \tilde{y}\cos\phi \tag{5.7}$$

Now we can express the derivatives of u and v with respect to $\tilde{x}$ and $\tilde{y}$ in terms of the derivatives u_x, v_x, u_y, v_y with respect to x and y. Then we see immediately that, if the C-R conditions are satisfied by the latter derivatives, they are satisfied also by the new derivatives. Thus the limit in Equation 5.3 is the same also for h approaching zero parallel to the new real and imaginary axes. If we take the further step of using, say, x and $\tilde{x}$ instead of x and y as independent variables in Equation 5.4, we can show also that the new and old real axes are equivalent in the same way, that the derivative is the same whether h approaches zero along one of these directions or the other. The upshot of these arguments is that the derivative is unique for *all* directions from which h may approach zero, provided only that the C-R conditions are satisfied with reference to *one* set of x and y axes in the z-plane. Thus these conditions are necessary *and* sufficient for the existence of the derivative $w'(z)$ at a given point z.[2]

Let us now inquire into the conditions for the existence (uniqueness) of a second derivative of $w(z)$ with respect to z. We thus inquire whether the real and imaginary parts of w' satisfy the C-R conditions, assuming that w' itself exists (i.e., that Equation 5.4 is satisfied). Let $w' = U + iV$; Equation 5.4 now tells us that

$$\begin{aligned} U &= u_x = v_y; \\ V &= v_x = -u_y. \end{aligned} \tag{5.8}$$

Thus we can get two expressions for each of the partial derivatives of U and V:

$$\begin{aligned} U_x &= u_{xx} = v_{yx}; \qquad V_x = v_{xx} = -u_{yx}; \\ U_y &= u_{xy} = v_{yy}; \qquad V_y = v_{xy} = -u_{yy}. \end{aligned} \tag{5.9}$$

Because the second "cross-" derivatives are equal, we can see that U and V satisfy the C-R conditions as a consequence of u and v

satisfying them (and of the existence of all the second partial derivatives of u and v). Thus, if the real and imaginary parts of a function $w(z)$ have all their second partial derivatives with respect to the real and imaginary parts of z, and if $w(z)$ has a unique first derivative $w'(z)$ with respect to z at a given point, it follows that it also has a unique second derivative $w''(z)$ at that point. This argument clearly can be extended to nth derivatives $w^{(n)}(z)$, provided u and v have all their partial derivatives through nth order. If u and v are *analytic* functions of x and y in a given region, then, and w' exists in that region, it follows that $w(z)$ is *analytic* (that is, has all derivatives) at all points in the region. In the absence of counter-indications, we shall in what follows assume that u and v are analytic wherever they satisfy the C-R conditions.

In fact, ordinarily it is safe to assume that a function f which is an analytic function of a real variable, $f(x)$, is also analytic when expressed as a function $f(z)$ of the complex variable z. If such a function has an obvious singularity, usually it can be taken for granted that the function is analytic at other points. If in doubt, one can explicitly apply the C-R equations as a test. There are some simple functions that are not analytic anywhere. Two examples are the complex conjugate, z^*, and the absolute value of z, $|z| = \sqrt{x^2 + y^2}$.

In case it may sometimes be more convenient to apply the C-R conditions to a function expressed in polar form, $w = u + iv = Re^{i\phi}$, or using an independent variable that is so expressed, $z = x + iy = re^{i\phi}$, we rewrite the conditions below in alternative forms; as usual, subscripts denote partial derivatives:

$$\begin{cases} 2\sin\theta\cos\theta\, v_\theta = -ru_r; \\ 2\sin\theta\cos\theta\, u_\theta = rv_r. \end{cases} \tag{5.6'}$$

$$\begin{cases} R_y = -R\phi_x; \\ R_x = R\phi_y. \end{cases} \tag{5.6''}$$

$$\begin{cases} 2R\sin\theta\cos\theta\, \phi_\theta = -rR_r; \\ 2\sin\theta\cos\theta\, R_\theta = rR\phi_r. \end{cases} \tag{5.6'''}$$

A by-product of the foregoing argument that follows also from Equation 5.9:

$$u_{xx} + u_{yy} = v_{xx} + v_{yy} = 0; \tag{5.10}$$

i.e., the real and imaginary parts of an analytic function of z both satisfy Laplace's equation in two dimensions. This property of analytic functions of z has been used as an aid to the solution of electrostatic and other problems in two dimensions. Each analytic function solves two such physical problems (one in which u is the potential, one in which v is). In fact, for reasons that follow from our next argument, the curves of constnat u are orthogonal to the curves of constant v (the two families of curves cross each other at right angles), so whichever family of curves are taken as equipotential curves, the other family are the electric field lines (lines of force). This method of solving Laplace's equation is known as the method of *conformal mapping.*

The term *conformal mapping* means "mapping with angles unchanged." The mapping of the z-plane into the w-plane defined by an analytic function $w(z)$ has the property of leaving invariant the angles at which curves cross. Figure 5.1 shows three points in the z-plane and their images in the w-plane, $w_1 = w(z_1)$, $w_2 = w(z_2)$, $w_3 = w(z_3)$.

Now we know that, as z_2 and z_3 approach z_1, the relations

$$\begin{aligned} w_2 - w_1 &= w'(z_1)(z_2 - z_1) + \text{higher-order terms}, \\ w_3 - w_1 &= w'(z_1)(z_3 - z_1) + \text{higher-order terms}, \end{aligned} \tag{5.11}$$

lose their higher-order (in z-differences) terms, and in the limit

$$\frac{w_2 - w_1}{w_3 - w_1} = \frac{z_2 - z_1}{z_3 - z_1}, \tag{5.12}$$

because of the uniqueness of $w'(z_1)$. If two complex numbers are expressed in polar form and then one is divided into the other, one finds that the phase angle of the quotient is the difference of phase angles of the numerator and the denominator. Equation 5.12 equates two such quotients of complex numbers, and thus implies that the difference of phase angles of numerator and denominator is the same for the two quotients. But this difference is just the angle marked ϕ in the diagram, which is thus the same in the two planes, in the limiting case in which the triangle approaches zero size. Thus the mapping is conformal.

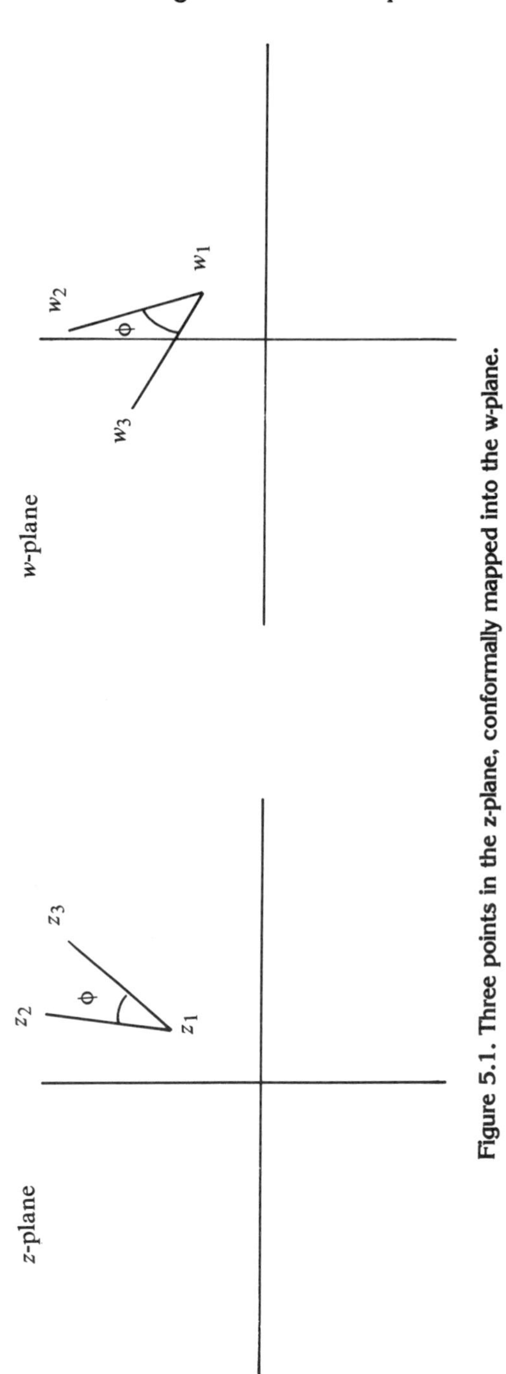

Figure 5.1. Three points in the z-plane, conformally mapped into the w-plane.

Having the Cauchy-Riemann conditions for analyticity, we are in a position to prove a very important theorem, *Cauchy's theorem,* concerning line integrals of analytic functions. Let C be a closed curve in the z-plane, and let $f(z)$ be a function of z that is analytic at every point within and on the curve C. Then the integral of $f(z)$ around C equals zero. In order to prove this result, we write

$$\oint_c f(z)\,dz = \oint_c [u(x,y) + iv(x,y)]\,(dx + idy)$$
$$= \oint_c (udx - vdy) + i \oint_c (vdx + udy). \tag{5.13}$$

In general, a differential form $Fdx + Gdy$ is exact, i.e., the differential of a function H of x and y, if and only if the partial derivatives F_y and G_x are equal, for if H exists these are its two second cross-derivatives, which have to be equal. Thus H can be constructed as a line integral of dH which is $Fdx + Gdy$. This criterion of exactness has a limitation: Although H exists in a region in which $F_y = G_x$, it may be multiple-valued. In general, this is so in a region in which the condition of exactness, $F_y = G_x$, fails at one or more points. In such a region, $Fdx + Gdy$, if integrated around a closed curve enclosing such a point at which $F_y \neq G_x$, does not integrate to zero. However, if the condition of exactness is satisfied at *every* point within and on the closed curve around which one integrates, the integral is zero:

$$\oint (Fdx + Gdy) = \oint dH = 0. \tag{5.14}$$

The two integrals in Equation 5.13 thus have to be zero, by virtue of the C-R conditions, which are the conditions of exactness of the two differential forms, and which have been assumed to hold at every point within and on C. Thus Cauchy's theorem is established.

The case in which the C-R relations fail at one or more points within C is a very important one, in which the integral of $f(z)dz$

around C may not vanish. In order to deal with this problem, we must first discuss Taylor series. If $f(z)$ is analytic at a point z_0, it can be expanded in Taylor series around z_0:

$$f(z) = f(z_0) + (z - z_0)f'(z_0) + \tfrac{1}{2}(z - z_0)^2 f''(z_0) + \ldots + (z - z_0)^n f^{(n)}(z_0)/ n! + \ldots \quad (5.15)$$

The existence of the terms in the series is guaranteed by the analyticity of f at z_0, and the convergence of the series to $f(z)$ at points sufficiently close to z_0 is guaranteed by the ratio test. This test, applied to the series of absolute values of the terms above, shows that there is a value of $|z - z_0|$, known as the radius of convergence of the series, such that within a circle of this radius, with center at z_0, the series of absolute values converges (i.e., the series converges absolutely), whereas outside of this circle the series of absolute values and also the original series both diverge. On the circle, the series converges at some points and diverges at others. In general, the radius of convergence of a Taylor series about a point z_0 is the distance from z_0 of the nearest point of nonanalyticity of the function. The points of nonanalyticity that will most interest us are the so-called isolated *poles*.

We can exemplify isolated poles by defining a function $g(z)$ that has one:

$$g(z) = f(z)/(z - z_0)^k. \quad (5.16)$$

Here we are taking $f(z)$ to be analytic throughout a region that contains z_0. We intend k to be a positive integer, so $g(z)$ is undefined (singular) at z_0, though one can show by the C-R conditions that g is analytic everywhere else that f is analytic.

Using a Taylor series to represent $f(z)$, we can express $g(z)$ also as a series of powers of $z - z_0$:

$$g(z) = f(z_0)/(z - z_0)^k + f'(z_0)/(z - z_0)^{k-1} + \ldots + f^{(k-1)}(z_0)/(k - 1)!\,(z - z_0) + \ldots + (z - z_0)^{n-k} f^{(n)}(z_0)/n! + \ldots \quad (5.17)$$

Such a power series, containing negative as well as positive powers of $z - z_0$, is known as a *Laurent series.* Taylor series are regarded as special cases of Laurent series, but a series containing negative powers is not a Taylor series. The series above converges exactly where the Taylor series for $f(z)$ converges, except the one point z_0, which is known as the position of a *pole of $g(z)$ of kth order;* the order is the power in the denominator of the leading term.

If $g(z)$ is expanded in power series around some point z_1 that is less than halfway from z_0 out to the circle of convergence of the series 5.17, such a point is one at which g is analytic, so the series is a Taylor series. The radius of convergence of this series is the distance from z_1 to the pole at z_0. Clearly this Taylor series can not converge at z_0, for g is singular there; nor can it converge at greater distances from z_1.

Returning to the series 5.17, expanded about z_0, let us seek to integrate $g(z)$ around a circle with its center at z_0 and a radius less than the radius of convergence of the series. On such a circle the series converges uniformly, so it can be integrated term by term. Letting $z - z_0 = re^{i\theta}$, we can set $dz = e^{i\theta}\,dr + ire^{i\theta}\,d\theta$. On the circle around which we wish to integrate, $dr = 0$. Thus the integral can be written as

$$\oint g(z)\,dz = \int_0^{2\pi} ire^{i\theta}\,d\theta[f(z_0)r^{-k}e^{-ik\,\theta} + f'(z_0)r^{-k+1}\,e^{-i(k-1)\theta} + \ldots + f^{(k-1)}(z_0)/(k-1)!\,re^{i\theta} + \ldots]. \quad (5.18)$$

The θ-dependent part of each term is a periodic function of θ with mean value zero, except the term in $(z - z_0)^{-1}$; in that one term the $e^{i\theta}$ in front of the brackets cancels the one in the denominator of the term, so there is no θ dependence left. Thus only that one term contributes to the integral. Its contribution is easily found:

$$\oint g(z)\,dz = 2\pi i f^{(k-1)}(z_0)/(k-1)! = 2\pi i\,a_{-1}. \quad (5.19)$$

The last of these forms refers to the writing of the Laurent series for $g(z)$ in the form

$$g(z) = a_{-k}(z - z_0)^{-k+1} + \ldots + a_{-1}(z - z_0)^{-1} + a_0 + a_1(z - z_0) + \ldots = \sum_{n=-k}^{\infty} a_n(z - z_0)^n. \qquad (5.20)$$

The integral in 5.19 arises entirely from the -1st power of $z - z_0$; it is proportional to the coefficient a_{-1}, which is called the *residue* of $g(z)$ at the pole at z_0. In case $g(z)$ is expressed as a function $f(z)$, analytic in the neighborhood of z_0, divided by a power of $z - z_0$, this coefficient a_{-1} has the value given in Equation 5.19. If $g(z)$ is not encountered in this form, one can still discover the function f by multiplying g by $(z - z_0)^k$; having found f, one can then calculate the appropriate coefficient in the Taylor expansion of f about z_0.

The theorems and devices by which one uses Equation 5.19 to evaluate integrals are known collectively as the *theory of residues.* The closed curve C is known as a *contour* and the process of integrating around such a contour is known as *contour integration*. The contour C was taken to be a circle because it was easy to integrate the series for g around a circle. However, according to Cauchy's theorem, C can be deformed into other shapes without affecting the integral, provided the deformation of the contour does not make it enclose other points of nonanalyticity of g, or exclude the pole at z_0. Such deformation has no effect because the region between the old and new contours is one in which $g(z)$ is everywhere analytic; the difference between the new and old integrals is an integral around this region of analyticity, and hence equals zero.

In case one wishes to integrate (z) around a region in which it has two or more poles, one can use Cauchy's theorem to express the integral as the sum of integrals taken around the separate poles. Each of these integrals can then be expressed as $2\pi i$ times a residue a_{-1} in a Laurent series for $g(z)$ about the appropriate pole. It is necessary to remember that each of these series is a *different* series, with a *different* a_{-1}. Commonly one can recognize by inspection what part of the integrand plays the part of the analytic

$f(z)$ at each of the poles of g, and the relevant term in the Taylor series for *this* f about *this* pole of g can be inferred. Having determined the various coefficients a_{-1}, we must then add them. Incidentally, the result 5.19 applies to integration in the positive (counterclockwise) direction around a contour. Integrating in the clockwise direction, we get the same result with a reversed sign.

Points of nonanalyticity of a function that is analytic almost everywhere are known generically as *singularities* of the function. Poles are singularities of one type. A pole of infinite order, having a Laurent series in which negative powers go to $-\infty$ (for example, the series for $e^{1/z}$ expanded around the origin), is called an *essential singularity*. The theory of residues applies to integrals around isolated essential singularities as it does to those around isolated poles of finite order.

There are singularities of another type, however, which require special treatment in contour integration. Although we shall not need to deal with these cases, we should say enough about them to prevent possible blunders in other problems not discussed in this book. The singularities in question are known as *branch points*. They are defined as points at which a function which has n values in the neighborhood has fewer than n distinct values. They are often discussed in geometric terms: the graph of a function of a complex variable is spoken of as a *Riemann surface*—a surface in a four-dimensional space, the graph of two real functions against two real variables. A multiple-valued function has a Riemann surface of two or more *sheets*—at each point z the function $f(z)$ has two or more values. In these terms, a branch point is a point of z at which two or more sheets of the Riemann surface merge. The importance of branch points to our present concerns is that, in going once around a branch point, we do not in general return to the value of the function $f(z)$ at which we started out; we return to the initial z but not to the initial value of $f(z)$; we are on a different sheet of the Riemann surface from the one on which we began the circuit. Then the argument that led to Equation 5.19 is no longer correct. In order to apply the method of contour integration to multiple-valued functions, we must either refrain from encircling any branch points or encircle two or more in such a way as to return to the initial values of both z

and $f(z)$. We shall not discuss these techniques further, for we shall not need to use them, but shall briefly discuss a simple example of a multiple-valued function which will recur in a later chapter.

Let $f(z)$ be the nth root of z, where n is a positive integer:

$$w = f(z) = z^{1/n}. \tag{5.21}$$

Determining w requires solving the equation

$$w^n = z, \tag{5.22}$$

which equation has n roots $s_j, j = 1, 2, \ldots, n$, so the equation can be written

$$(w - s_1)(w - s_2)(w - s_3) \ \ldots \ (w - s_n) = w^n - z = 0. \tag{5.23}$$

This form permits various inferences, e.g., that the sum of the nth roots of any complex number is zero, but in this case it is easy to calculate the roots explicitly. Express z in polar form:

$$z = re^{i\theta}. \tag{5.24}$$

Then

$$w = z^{1/n} = r^{1/n} e^{i\theta/n}. \tag{5.25}$$

The nth root of r can be taken to be real and nonnegative, so it can serve as the absolute value of w; w is an n-valued function of z because the θ that enters z can be chosen in n different ways that do not affect z but do affect w. The phase angle of a complex number can be increased by any multiple of 2π without changing the number, because $e^{2\pi i} = 1$. Thus, taking θ to lie between 0 and 2π, we can write

$$z = r\, e^{i(\theta + 2\pi j)}, \tag{5.26}$$

where j is any integer. Now we have

$$w = z^{1/n} = r^{1/n} e^{i(\theta + 2\pi j)/n}. \tag{5.27}$$

As j assumes successive integer values from 1 to n, z is the same number every time, but w moves around a circle of radius $r^{1/n}$, in equal increments of phase angle; the angle between successive values of w is $2\pi/n$, and there are n distinct values before they begin repeating. Thus w is an n-valued function of z for all values of z except zero; zero is a branch point of the function $z^{1/n}$, at which all the roots merge into the common value zero.

An increase of θ by 2π produces one trip around the origin in the z-plane, but a trip only one nth of the way around the origin in the w-plane. Thus, if integrating w in a circle once around the origin in the z-plane, one would not really have returned to the starting point; if the integrand contained, with w, a factor that had a pole somewhere within the contour being contemplated as a path of integration, one should try to change the path so it would go around the pole but not around the branch point at the origin.

NOTES

1. Among the well-known books on functions of a complex variable are: Churchill, R. V., J. W. Brown, and R. F. Verhey. *Complex Variables and Applications.* New York; McGraw-Hill, 1974; Franklin, P. 3d ed. *Functions of Complex Variables.* Englewood Cliffs, NJ: Prentice-Hall, 1947.
2. The uniqueness of the derivative at a point must survive also the use of various complicated spirals along which h may approach zero. In fact, the C-R conditions and the continuity of the functions u and v are sufficient to guarantee uniqueness. See Churchill, R. V., and J. W. Brown. *Fourier Series and Boundary Value Problems.* 3d ed. New York: McGraw-Hill, 1978.

PROBLEMS

5.1. If $f(z)$ and $g(z)$ are both analytic at every point in a region R of the z-plane, determine where, in this region, each of the following functions is analytic:

(a) $f + g$
(b) $1/f$
(c) fg
(d) $g^*/|g|^2$
(e) f^*
(f) $|g|^2$
(g) $g^*/f|g|^2$
(h) $(1 + f)f^*$
(i) f/g
(j) $(f + f^*)/(1 + |f|)$
(k) fe^g
(l) $(g - f)^{1/2}$

5.2. Integrate each of the following functions, without using the theory of residues, counterclockwise around the square $(1 + i, -1 + i, -1 - i, 1 - i)$. Then check your answers for consistency with the theory of residues. In ambiguous cases, specify in which quadrant each quantity lies which is obtained by integration along one side of the square.

(a) z
(b) z^2
(c) $1/z$
(d) $z^{-1/2}$
(e) e^z
(f) e^{iz}

6

Evaluation of Certain Fourier Integrals—Causality, Green's Functions

The immediate purpose of the foregoing chapter was the evaluation of integrals like that in Equation 4.53. In the present chapter we continue this process by applying the theory of residues to such integrals. We rewrite the integral here:

$$x(t) = (2\pi)^{-1/2} \int_{-\infty}^{\infty} \frac{-\alpha(\omega)\, e^{i\omega t}\, d\omega}{(\omega - \omega_1)\,(\omega - \omega_2)}, \tag{6.1}$$

where we have now factored the denominator. The roots, ω_1 and ω_2, are, for an underdamped oscillator:

$$\begin{aligned} \omega_1 &= i\gamma + \sqrt{\omega_0^2 - \gamma^2} = i\gamma + \omega' \\ \omega_2 &= i\gamma - \sqrt{\omega_0^2 - \gamma^2} = i\gamma - \omega'. \end{aligned} \tag{6.2}$$

Aside from the unspecified $\alpha(\omega)$, this integrand is analytic everywhere except the two points ω_1 and ω_2, which are poles of first order. What other poles there may be depends on $\alpha(\omega)$.

We shall evaluate such integrals by means of the theory of residues, using a contour that is closed by an infinite semicircle either above or below the real axis. More precisely, we shall treat the improper integral 6.1 as the limit of an integral from $-R$ to R, as R goes to infinity. For each value of R, we can define the contour $C_+(R)$ consisting of the interval from $-R$ to R, plus a semicircle $S_+(R)$ which joins the limits $-R$ and R above the real axis. Also, for each value of R, we can define the closed contour $C_-(R)$, consisting of the interval from $-R$ to R, plus the semicircle $S_-(R)$ which joins the points at $-R$ and R below the real axis. These paths of integration are shown in Figure 6.1, along with the points ω_1 and ω_2, for a value of R whereby the poles are inside the upper contour C_+.

Now, representing integrals schematically, we can write

$$\int_{-R}^{R} + \int_{S_+(R)} = \oint_{C_+(R)} \quad \text{and} \quad \int_{-R}^{R} + \int_{S_-(R)} = \oint_{C_-(R)}, \tag{6.3}$$

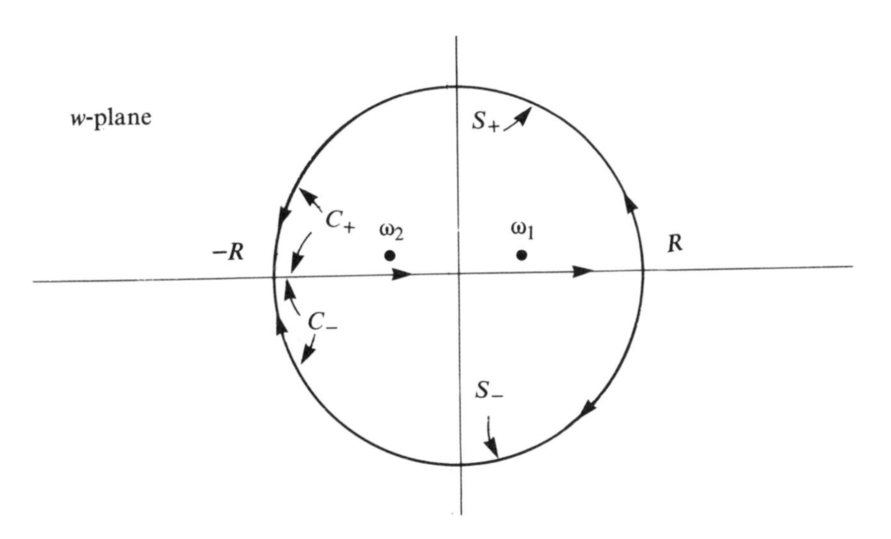

Figure 6.1. Paths in the complex plane for evaluation of Fourier integrals, before the limits have gone to infinity.

provided all the indicated integrals exist. We shall show that, in every case of interest, the integral over either S_+ or S_- goes to zero as R goes to infinity. Thus we can equate the improper integral 6.1 to either the contour integral around the infinite C_+ or that around the infinite C_- (which is taken in the negative direction).

In order to discuss the integrals over S_+ and S_-, we write the integrand of 6.1 as

$$g(\omega) = A(\omega)\, e^{iT\omega} = A(\omega)\, e^{iTR(\cos\theta + i\sin\theta)}, \tag{6.4}$$

where in the last step we have written ω in polar form appropriate to points on $S_+(R)$ and $S_-(R)$. $A(\omega)$ is defined as an algebraic function which includes the algebraic denominator in 6.1 and also any algebraic part of $\alpha(\omega)$. T is the time t plus any additional contribution that may come from an exponential factor in $\alpha(\omega)$. The form 6.4, though not completely general, is general enough for our present purposes. Now, writing the integrand $g(\omega)$ as

$$g(\omega) = A(\omega)\, e^{iTR\cos\theta}\, e^{-TR\sin\theta}, \tag{6.5}$$

we see that the behavior as $R \to \infty$ depends on the sign of $T\sin\theta$, for the last factor will determine whether the absolute value of g goes to 0 or ∞; A is an algebraic function, and the second factor has absolute value one. Thus, in seeking a semicircle on which g vanishes at $R = \infty$, we choose S_+ if $T > 0$ and S_- if $T < 0$.

Even so, however, we still have to prove that no trouble arises from the ends of the semicircle, where $\sin\theta = 0$. For definiteness we shall examine the case of S_+, $T > 0$; the other case involves similar reasoning. Recalling that $d\omega = e^{i\theta}\, dr + ire^{i\theta}\, d\theta$, we have

$$\int_{S_+} g(\omega) d\omega = \int_0^\pi iRA(\omega)\, e^{i\theta}\, e^{iTR\cos\theta}\, e^{-TR\sin\theta}\, d\theta \equiv \int_0^\pi U_R(\theta) d\theta. \tag{6.6}$$

We shall show that this integral vanishes as $R \to \infty$, subject to a weak condition on A, by showing that the absolute value of the integrand U_R yields an integral that goes to zero; this establishes that the original integral goes to zero. Thus,

$$\int_0^\pi |U_R(\theta)|\, d\omega =$$
$$\int_0^\pi R\,|A(\omega)|\, e^{-TR\sin\theta}\, d\theta \leqslant \int_0^\pi R\,|A(\omega)|\, e^{-TRL(\theta)}\, d\theta, \tag{6.7}$$

where $L(\theta) \leqslant \sin\theta$ for $0 \leqslant \theta \leqslant \pi$, $L(\theta)$ being defined as

$$L(\theta) = 2\theta/\pi \text{ for } 0 \leqslant \theta \leqslant \tfrac{1}{2}\pi$$
$$L(\theta) = 2(1 - \theta/\pi) \text{ for } \tfrac{1}{2}\pi < \theta \leqslant \pi. \tag{6.8}$$

Now, letting $Q(R)$ be the largest value of $|A(\omega)|$ on S_+, we can argue that

$$\int_0^\pi |U_R(\theta)|\, d\theta < \pi\,[RQ(R)/-TR][e^{-TR} - 1 - 1 + e^{-TR}]/2$$
$$\rightarrow (\pi/TR)\, RQ(R) = \pi Q(R)/T. \tag{6.9}$$

Thus we see that the original integral around S_+ has an absolute value less than or equal to π times the maximum absolute value of $A(\omega)$ on S_+, divided by T, in the limit as $R \rightarrow \infty$. So a sufficient condition for the integral on S_+ to go to zero for $T > 0$ (and the one on S_- to go to zero for $T < 0$) is that $|A(\omega)| \rightarrow 0$ as $|\omega| \rightarrow \infty$, uniformly (i.e., in all directions).

Then, depending on the sign of T, we have only to verify this weak condition on A in order to express Fourier integrals like 6.1 as contour integrals around C_+ and C_-. In each case, assuming a single-valued integrand, the Fourier integral can be evaluated by the method of residues. The condition on A is very likely to be true, since the part of A that does not depend on $\alpha(\omega)$ already acts like ω^{-2} at infinity, so $\alpha(\omega)$ can even go to infinity like $\omega^{2-\varepsilon}$ without violating the condition.

In order to demonstrate the method, let us ask for the response of an underdamped oscillator to the applied force

$$F(t) = 0 \qquad (\text{for } t < t_0),$$
$$F(t) = mKe^{-b(t-t_0)} \qquad (\text{for } t \geqslant t_0). \tag{6.10}$$

Then $\alpha(\omega)$, the transform of F/m, is (cf. Equation 4.29)

$$\alpha(\omega) = -e^{-i\omega t_0}\,Ki/(\omega - ib)\,\sqrt{2\pi}. \tag{6.11}$$

Before attacking the more complicated problem of calculating $x(t)$, let us apply our method to this $\alpha(\omega)$ itself, to see whether we can recover $f(t) = F(t)/m$ by doing the inverse Fourier integral of α:

$$\begin{aligned} f(t) &= (2\pi)^{-1/2} \int_{-\infty}^{\infty} \alpha(\omega)\, e^{i\omega t}\, d\omega \\ &= (K/2\pi i) \int_{-\infty}^{\infty} e^{i\omega(t-t_0)}\, d\omega/(\omega - ib). \end{aligned} \tag{6.12}$$

Here the quantity T is $t - t_0$, and $A(\omega) = (\omega - ib)^{-1}$ which does go to zero at $\omega = \infty$. Then for $t < t_0$ we equate the integral 6.12 to that around C_-. The latter encircles no poles, so the integral is zero:

$$f(t) = 0 \qquad \text{for } t < t_0. \tag{6.13}$$

For $t > t_0$ we integrate around C_+, thus encircling the one pole at ib. This being a pole of first order, the residue is simply the analytic part of the integrand evaluated at $\omega = ib$:

$$f(t) = 2\pi i(K/2\pi i)\, e^{i(ib)(t - t_0)} = Ke^{-b(t-t_0)} \text{ (for } t > t_0). \tag{6.14}$$

So we have recovered the correct $f(t)$ from $\alpha(\omega)$, thus justifying the remarks following Equation 4.29. Although $\alpha(\omega)$ is a continuous function of ω at all points except ib, its Fourier integral equals a discontinuous function of t.

More complicated integrals, such as the one that yields $x(t)$ for this applied force, have this same property. Although $x(t)$ is not discontinuous, it does fail to be an analytic function of the real variable t at $t = t_0$—as it should, for a discontinuous force should produce a discontinuous x. This discontinuous behavior again follows from the choice between C_+ and C_-, which choice is again forced on us by the behavior of the integrand on S_+ and S_- which leads to one of these two integrals being zero.

The Fourier integral for $x(t)$ is

$$x(t) = -(K/2\pi i) \int_{-\infty}^{\infty} e^{i\omega(t-t_0)}\, d\omega/(\omega - ib)(\omega - \omega_1)(\omega - \omega_2), \tag{6.15}$$

where $A(\omega)$ is even better behaved at ∞ than before. The integrand is single-valued and has three poles in the upper half-plane. Again the sign of $t - t_0$ determines the choice of contour, C_+ or C_-. As before, when $t < t_0$, we integrate around C_-, encircling no poles, so

$$x(t) = 0 \text{ (for } t < t_0). \tag{6.16}$$

This reassuring fact, that the oscillator does not start to move before any force acts on it, is referred to under the rather pretentious name of *causality,* which in this context means merely that no effect precedes its cause—a property that is built into the definitions of the words "cause" and "effect," whereby we would not apply these words in speaking of phenomena that did not have the appropriate time order. Thus, what is being partly verified in Equation 6.16 is the correctness of the mathematical method, which *ought* to give the same result as would direct solution of the differential equation with application of initial conditions—and, so far, is doing so.

When $t > t_0$, we integrate around C_+, encircling three poles of the integrand. Each pole is of first order, so each residue equals the analytic (at that pole) part of the integrand evaluated at the pole. Thus, for $t > t_0$,

$$\begin{aligned} x(t) = -K\, [&e^{-b(t-t_0)}/(ib - \omega_1)(ib - \omega_2) \\ &+ e^{i\omega_1(t-t_0)}/(\omega_1 - ib)(\omega_1 - \omega_2) \\ &+ e^{i\omega_2(t-t_0)}/(\omega_2 - ib)(\omega_2 - \omega_1)], \end{aligned} \tag{6.17}$$

where we have added the three residues. The reader is invited to simplify this expression by finding a common denominator, etc., and, for example, thus confirming that it is real, that it satisfies the equation of motion with the assumed force acting, and that $x(0) = \dot{x}(0) = 0$.

Here we shall look further only at the special case $b = 0$, whereby the driving force is of step-function form. Setting $b = 0$, we get, for $t > t_0$,

$$x(t) = -K[1/\omega_1\omega_2 + e^{i\omega_1(t-t_0)}/\omega_1(\omega_1 - \omega_2) + e^{i\omega_2(t-t_0)}/\omega_2(\omega_2 - \omega_1)]. \tag{6.18}$$

Recalling that

$$\omega_1 = i\gamma + \omega',$$

and

$$\omega_2 = i\gamma - \omega', \tag{6.2'}$$

and that

$$\omega_0^2 = k/m = \omega'^2 + \gamma^2,$$

we can simplify Equation 6.18:

$$x(t) = \frac{Km}{m\omega_0^2}\left[1 - e^{-\gamma(t-t_0)}\left(\cos\omega'(t - t_0) + \times \frac{\gamma}{\omega'}\sin\omega'(t - t_0)\right)\right], \tag{6.19}$$

where the factor in front is so written as a reminder that the force $F(t)$ equals $mf(t)$ and therefore, (as in Equation 6.10) has a factor mK, and that $m\omega_0^2$ is the oscillator's force constant k. Thus we see x approaching the constant value F/k, as it should in the case of a constant force turned on at t_0. While it is approaching this value, it undergoes a damped oscillation which satisfies the homogeneous equation of motion and embodies the transient motion. Thus we have, in this case, verified our earlier claim (after Equation 4.53) that by adding together many steady-state responses of our oscillator we could discover its transient motion.

Let us now contemplate another example, which somewhat resembles that of Problem 4.6e.

$$f(t) = F(t)/m = \tfrac{1}{2}be^{-b(t-t_0)} \qquad (\text{for } t > t_0).$$
$$= \tfrac{1}{2}be^{b(t-t_0)} \qquad (\text{for } t < t_0). \tag{6.20}$$

Since the Fourier transform of a sum of terms is the sum of transforms of the separate terms, we can take part of $\alpha(\omega)$ to be given by Equation 6.11, with $K = \frac{1}{2}b$. Calculating the transform of the part of f before t_0 and adding it to what we have already, we get

$$\alpha(\omega) = \frac{bie^{-i\omega t_0}}{2\sqrt{2\pi}}\left(\frac{1}{\omega + ib} - \frac{1}{\omega - ib}\right). \tag{6.21}$$

The part of f acting before t_0 has now contributed a pole in the lower half-plane. The use of C_+ for $t > t_0$ and C_- for $t < t_0$ is still valid, as it was formerly, but $x(t)$ is different in both time periods from what it was previously. Now x is nonzero before t_0, as it should be; it arises from the pole at $-ib$

$$x(t) = \tfrac{1}{2}b\, e^{b(t-t_0)}/[\omega'^2 + (b + \gamma)^2] \quad (\text{for } t < t_0). \tag{6.22}$$

For $t > t_0$, integrating around C_+, one gets no residue from the pole at $-ib$, but the result 6.17 is not complete for this case, because the term in $1/(\omega + ib)$ will contribute to the residues at ω_1 and ω_2. Thus 6.17 becomes

$$\begin{aligned} x(t) = -\tfrac{1}{2}b[&e^{-b(t-t_0)}/(ib - \omega_1)(ib - \omega_2) \\ &+ e^{i\omega_1(t-t_0)}/(\omega_1 - ib)(\omega_1 - \omega_2) \\ &+ e^{i\omega_2(t-t_0)}/(\omega_2 - ib)(\omega_2 - \omega_1) \\ &- e^{i\omega_1(t-t_0)}/(\omega_1 + ib)(\omega_1 - \omega_2) \\ &- e^{i\omega_2(t-t_0)}/(\omega_2 + ib)(\omega_2 - \omega_1)] \quad (\text{for } t > t_0). \end{aligned} \tag{6.23}$$

Note that the force before t_0 has added the last two terms within the brackets to the motion after t_0.

We have chosen the applied force so that it would have unit area regardless of the (real, positive) value of b. Now let us look at the limit of $x(t)$ as $b \rightarrow \infty$. This process converts $f(t)$ to the sort of instantaneous impulse that we have called a delta function of $t - t_0$ (subject to certain interpretive understandings). Appropriately, Equation 6.22 shows that, because of the exponential factor, $x(t) = 0$ before t_0 in this limit.

After t_0, Equation 6.23 gives the limiting $x(t)$:

$$\begin{aligned} x(t) &= (2i)^{-1}[e^{i\omega_1(t-t_0)} - e^{i\omega_2(t-t_0)}]/\omega' \\ &= e^{-\gamma(t-t_0)}[\sin\omega'(t - t_0)]/\omega' \qquad (\text{for } t > t_0). \end{aligned} \tag{6.24}$$

In the same sense that Equations 6.16 and 6.19 give the response of the oscillator to a step function at t_0, Equations 6.22 (in the limiting case) and 6.24 give the response to a delta function at t_0. We obtained the step-function result by getting the response to a family of forces of which the step function was the limit as the parameter b went to zero. We could *not* have done the step-function problem directly by our present method; the step function has no Fourier transform. If one assigns it a transform by letting b go to zero in the transform of the exponential, this "transform" has a pole on the real axis, so one is in doubt as to whether it is within C_+ or C_-. The method used above, though, is perfectly sound, despite being indirect. We can argue that, even if $b = 10^{-90}$ sec^{-1}, the transform of the force exists and has a pole above the real axis. No one could ever tell by observation whether a step-function force is quite a step function or an exponential with a very tiny value of b. If 10^{-90} is not small enough, let $b = 10^{-900}$. It is inescapable that the response to a true step function is the limit, as $b \rightarrow 0$, of the responses to the family of exponentials. Thus, even though the step function is a legitimate function whose definition entails no contradictions, it can be Fourier-analyzed only by indirection.

In contrast, as we have seen in Chapter 4, there is no such thing as a delta function; nevertheless, our indirect (and legitimate) procedure for using it as force/mass was unnecessary. As we have suggested in the vicinity of Equation 4.44 the delta function can be used in calculations provided one is watchful for anomalies. Equation 4.44, giving the Fourier transform of $\delta(t)$, is a special case of Equation 4.45. Applying that Equation 4.45 to $\delta(t - t_0)$, we get $\alpha(\omega) = (2\pi)^{-1/2} e^{-i\omega t_0}$. Then, substituting into Equation 6.1, we get

$$x(t) = -(2\pi)^{-1} \int_{-\infty}^{\infty} d\omega e^{i\omega(t-t_0)}/(\omega - \omega_1)(\omega - \omega_2). \qquad (6.25)$$

The poles being above the real axis, we can again argue that $x(t) = 0$ for $t < t_0$. When $t > t_0$, $x(t)$ arises from the residues at ω_1 and ω_2:

$$x(t) = [e^{-\gamma(t-t_0)} \sin\omega'(t - t_0)]/\omega' \qquad (\text{for } t > t_0). \qquad (6.24')$$

This result agrees with the earlier one obtained by virtuous indirection. It appears that, even if the delta function does not exist, one can get away with pretending that it does.

What, then, does this result mean? It cannot represent the response to an impulse m at the instant t_0, for a finite impulse cannot be instantaneous. Instead, it represents the limit of responses to an impulse, m, centered about the time t_0, as it becomes more and more concentrated in time. As far as observation is concerned, an impulse lasting 10^{-113} second may as well be instantaneous—yet even an impulse that brief incurs no mathematical contradictions. To study the effect of such a brief impulse as a physicist, one may as well go to the limit of zero duration (that is, use a delta function).

Let us now use two different arguments, one legitimate and one illegitimate, to derive another method for solving oscillator problems. Equation 4.45 can be restated as

$$f(t) = \int_{-\infty}^{\infty} \delta(t - t')f(t')dt'. \qquad (6.26)$$

This relationship can be construed as being, like the Fourier integral, an expansion of a function $f(t)$ as a (continuous) superposition of orthogonal functions. In the Fourier case the functions are the $e^{i\omega t}$, with the parameter ω ranging from $-\infty$ to ∞, whereas in the present case the orthogonal functions are the

$\delta(t - t')$, with the parameter t' ranging from $-\infty$ to ∞. Our use of Fourier analysis for determining the motion of an oscillator depended on the argument that the response to a sum of $e^{i\omega t}$ forces is the sum of responses to the separate forces. Let us, then, use the same argument on the expression 6.26: the response of the oscillator to a sum of delta-function forces is the sum of the responses to the separate forces. Let us denote by $x_{t'}(t)$ the response, deducible from Equation 6.24 and accompanying remarks, to the force/mass $f_{t'}(t) = \delta(t - t')$:

$$
\begin{aligned}
x_{t'}(t) &= 0 \text{ for } t < t', \\
&= (1/\omega')e^{-\gamma(t-t')}\sin\omega m'(t - t') \qquad (\text{for } t \leqslant t').
\end{aligned}
\tag{6.24''}
$$

Then the response to the entire sum 6.26 should be

$$
\begin{aligned}
x(t) &= \int_{-\infty}^{\infty} f(t')\, x_{t'}(t)\, dt' \\
&= (1/\omega') \int_{-\infty}^{t} f(t')\, e^{-\gamma(t-t')}\sin\omega'(t - t')\, dt',
\end{aligned}
\tag{6.27}
$$

where the upper limit on the last integral is t instead of ∞ because $x_{t'}(t) = 0$ for $t' > t$. Physically, this means that, again, effects follow causes; forces acting at times after t do not affect the motion at time t.

The function $x_{t'}(t)$ is called a *Green's function,* and is usually written $G(t - t')$, so Equation 6.24'' defines the oscillator's Green's function. Note that the Green's function is indeed a property of the oscillator and does not depend at all on the applied force. In fact, $G(t - t')$, in its dependence on t, is a solution of the homogeneous equation of motion, in which there is no applied force. The solution 6.24'', the response of the system to a delta-function force, which *is* the Green's function, is that solution $x_H(t)$ which satisfies $x(t') = 0$, $m\dot{x}(t'+) = m$ (because the impulse at t' equalled m). So Equation 6.27 (rewritten below in more customary notation) presents the motion in a form in which the contribution of the force is separated from that of the system:

$$x(t) = \int_{-\infty}^{t} G(t - t')\,f(t')\,dt';$$

$$G(t - t') = 0 \qquad (\text{for } t' > t), \tag{6.28}$$

$$= \frac{1}{\omega'} e^{-\gamma(t-t')} \sin\omega'(t - t'), \qquad (\text{for } t' < t).$$

Having derived these relationships by means of such dubious devices as delta functions, let us now proceed more legitimately. In Equation 6.1, let us substitute for $\alpha(\omega)$ the Fourier integral expression for it in terms of $f(t)$—but we must use t' instead of t in that integral because the whole expression is a function of t, whereas t' gets integrated out. Thus

$$x(t) = -(2\pi)^{-1} \int_{-\infty}^{\infty} \frac{d\omega e^{i\omega t}}{(\omega - \omega_1)(\omega - \omega_2)} \int_{-\infty}^{\infty} f(t') e^{-i\omega t'} dt'$$

$$= -(2\pi)^{-1} \int_{-\infty}^{\infty} f(t') dt' \int_{-\infty}^{\infty} \frac{e^{i\omega(t - t')} d\omega}{(\omega - \omega_1)(\omega - \omega_2)} \tag{6.29}$$

$$= \int_{-\infty}^{\infty} G(t - t')\,f(t')\,dt',$$

where

$$G(t - t') = -(2\pi)^{-1} \int_{-\infty}^{\infty} e^{i\omega(t-t')}\,d\omega/(\omega - \omega_1)(\omega - \omega_2), \tag{6.25'}$$

which is just the response to a delta function as already given in Equation 6.25. Equation 6.1 shows the integrand of the Fourier integral for $x(t)$ as a product, where the factor $\alpha(\omega)$ is the part coming from the applied force, and the rest depends on the properties of the oscillator. The Green's function is the transform of this "rest," and the transition from the form 6.1 to the form 6.29 can be achieved as an application of the convolution theorem: the transform of a product 6.1 is the convolution 6.29 of the transforms (times a constant factor).

Often the Green's function provides the easiest way to calculate the response of an oscillator to a given $f(t)$. It has the advantage over the Fourier integral that problems of convergence seldom arise. As simple as it looks, however, one can easily make mistakes by carelessness with notation. We shall illustrate the method by calculating the reponse of an underdamped oscillator to the force/mass

$$\begin{aligned} f(t) &= 0 \text{ (before } t_0 \text{ and after } t_1); \\ f(t) &= A \text{ (between } t_0 \text{ and } t_1). \end{aligned} \tag{6.30}$$

Now Equation 6.27 shows explicitly that $x(t) = 0$ before the earliest time at which a force acts. Then, between t_0 and t_1, we have

$$\begin{aligned} x(t) &= (A/\omega') \int_{t_0}^{t} e^{-\gamma(t-t')} \sin\omega'(t - t')dt' \\ &= \frac{A}{\omega_0^2} \left\{1 - e^{-\gamma(t-t_0)}\left[\cos\omega'(t - t_0) + \frac{\gamma}{\omega'} \sin\omega'(t - t_0)\right]\right\} \\ &\quad \text{(for } t_0 \leqslant t \leqslant t_1). \end{aligned} \tag{6.31}$$

After t_1 we have

$$\begin{aligned} x(t) &= (A/\omega') \int_{t_0}^{t_1} e^{-\gamma(t-t')} \sin\omega'(t - t')dt' \\ &= \frac{A}{\omega_0^2} \left\{1 - e^{-\gamma(t-t_1)}\left[\cos\omega'(t - t_1) + \frac{\gamma}{\omega'} \sin\omega'(t - t_1)\right]\right. \\ &\quad \left. -e^{-\gamma(t-t_0)}\left[\cos\omega' (t - t_0) + \frac{\gamma}{\omega'} \sin\omega' (t - t_0)\right]\right\} \\ &\quad \text{(for } t \geqslant t_1). \end{aligned} \tag{6.32}$$

A reader who has not previously carried through such calculations should carefully do this one, checking whenever possible with the expressions above.

There are two interesting limiting cases to look at. One is the limit $t_1 \to \infty$, which renders $f(t)$ a step function. Then the form 6.31 holds for all times later than t_0. It agrees with the response to a step function in Equation 6.19, as it should, but it was considerably easier to get; no limiting processes were needed, for even if $t_1 \neq \infty$, for $t < t_1$ the oscillator has no way of knowing that the force is not a step function; until t_1, $x(t)$ is precisely the response to a step function, as if the force were going to remain constant from then on, and this $x(t)$ was calculated in a straightforward way, with no difficulties of convergence.

The other interesting case is that in which the pulse has unit area (impulse = m), and t_1 approaches t_0. Then the period from t_0 to t_1 dwindles to nothing and the only nonzero response comes after t_1, as in equation 6.32. Letting $t_1 = t_0 + d$ and $A = 1/d$ in Equation 6.32, we treat $t - t_1$ as $t - t_0 - d$ in the exponential and trigonometric functions:

$$x(t) = \frac{e^{-\gamma(t-t_0)}}{\omega_0^2}\Big\{\cos\omega'(t-t_0)\,\frac{\omega'(e^{\gamma d}\cos\omega' d - 1) - \gamma e^{\gamma d}\sin\omega' d}{\omega' d}$$

$$+ \sin\omega'(t-t_0)\,\frac{\omega' e^{\gamma d}\sin\omega' d + \gamma(e^{\gamma d}\cos\omega' d - 1)}{\omega' d}\Big\}$$

$$(\text{for } t \geqslant t_0 + d). \tag{6.33}$$

Here we have used the identities for trigonometric functions of differences of angles. Next we let $d \to 0$, using power series for the exponential and trigonometric functions as an aid in working out the limit. The limiting form for $x(t)$, $t > t_0$, when the pulse has become a delta function, is

$$x(t) = [e^{-\gamma(t-t_0)}\sin\omega'(t-t_0)]/\omega', \tag{6.24''}$$

as we have seen it repeatedly before. In this case, as in previous cases, we would not have had to do the delta-function problem in such an elaborate way. We could have simply used the delta function for $f(t)$ in Equation 6.28, thus being reminded that $G(t - t')$ *is* the response to a delta function at t'.

NOTES

1. On the use of contours such as C_+ and C_-, see Smith, L. P. *Mathematical Methods for Scientists and Engineers* Englewood Cliffs, NJ. Prentice-Hall, Dover Reprint, 1961.
2. On Green's functions in general, see Morse, P. M., and H. Feshbach. *Methods of Theoretical Physics.* New York: McGraw Hill, 1953.

PROBLEMS

6.1. Calculate the following integrals:

$$\text{a)} \int_{-\infty}^{\infty} \frac{\cos x \, dx}{1 + x^2}, \qquad \text{b)} \int_{-\infty}^{\infty} \frac{dx}{1 + x^4}$$

6.2. Calculate the inverse of the Fourier integral of Problem 4.6e, verifying that you thus recover the original $f(t)$.

6.3. Calculate the response of an underdamped oscillator to the force $F(t) = 0$ for $t < 0$; $F(t) = Ae^{-bt} \sin ct$ for $t > 0$, using the Fourier integral directly, and using the Green's function.

6.4. Solve Problem 6.3 for the case in which $b = \gamma$ and $c = \omega'$.

6.5. Calculate the Green's function of a critically damped RLC series circuit, whereby the voltage across the resistance is given in terms of the emf applied across the whole series combination.

6.6. Apply the Green's function of Problem 6.5 to the case in which the applied emf is $F(t) = A$ for $t < 0$; $F(t) = 0$ for $t > 0$. Solve the same problem by direct use of a Fourier integral.

6.7. Calculate the response of an underdamped oscillator to the force $F(t) = 0$ for $t < 0$; $F(t) = Ate^{-bt}$ for $t > 0$. Use the Fourier integral and the Green's function, and compare the two solutions.

6.8. The emf applied across an underdamped series RLC circuit is $F(t) = A\delta(t - t_0)$. Calculate the time-dependent voltage across C; across L.

6.9. Use a Green's function to check Problem 1.9.

6.10. Calculate the response of an underdamped oscillator to the derivative of a delta function: the force is $F(t) = A\delta'(t - t_0)$. How is this response related to the Green's function $G(t - t_0)$?

6.11. Referring to Problem 3.7 work out the Green's function which gives the time-dependent position of a mass hanging on a coil spring, in terms of the time-dependent height of the spring's point of support.

7

Electrical Networks

We have now devoted six chapters to analyzing single oscillators (i.e., oscillators with one current loop or one mass) and have developed techniques for solving single-oscillator problems. In the present chapter we extend these techniques to networks, i.e., systems of interacting single oscillators. Chapters 9-13 analyze certain networks in a more systematic way, using new techniques. Most of the networks considered in this and later chapters are electrical rather than mechanical, because a much greater variety of electrical than mechanical networks can be constructed.

Let us then suppose that we wish to analyze the network shown in Figure 7.1, with all the applied emfs being at a common frequency ω. The round objects marked F_1, F_{14}, and so on, are voltage sources (generators), all at frequency ω, and having complex voltage amplitudes given by the Fs; the arrowheads near them indicate the positive sense of each voltage. Similarly, the Is are complex amplitudes of loop currents, whose positive sense is indicated by the curved arrows. By using loop instead of branch currents, we automatically satisfy the Kirchhoff law about the sum of currents at a point, and have only to choose the currents so as to satisfy the other Kirchhoff law, about the sum of emfs and voltage drops around a loop. The Zs are complex impedances, in general series combinations, because parallel and other combinations lead to the definition of more loops. So the general Z has the form $i\omega L + R + 1/i\omega C$. We are assuming that there are no mutual inductances, although these can be included that there are no mutual inducatances, although these can be included without much more trouble. The type of circuit depicted is general enough to include so-called active elements such as vacuum tubes and transistors; in these cases, the different generators have voltages that are related. We shall not try to analyze circuits containing such devices, but shall soon specialize to the case in which there is only one generator. As the circuit is drawn, however, the currents have to satisfy the following equations:

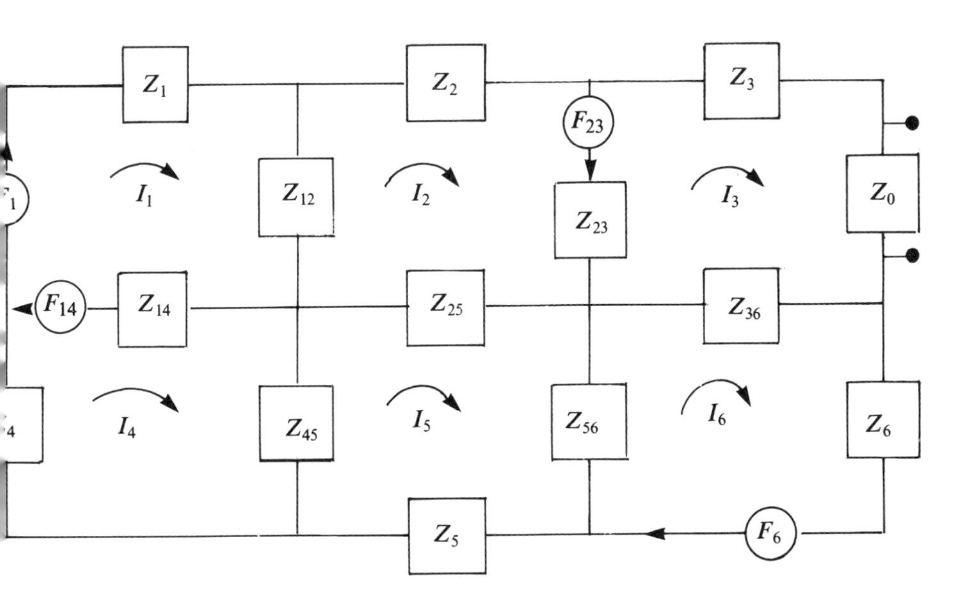

Figure 7.1. An electrical network.

$$
\begin{aligned}
F_1 + F_{14} &= I_1 (Z_1 + Z_{12} + Z_{14}) - I_2 Z_{12} - I_4 Z_{14} \\
F_{23} &= -I_1 Z_{12} + I_2(Z_{12} + Z_2 + Z_{23} + Z_{25}) \\
&\quad - I_3 Z_{23} - I_5 Z_{25} \\
-F_{23} &= -I_2 Z_{23} + I_3(Z_{23} + Z_3 + Z_0 + Z_{36}) - I_6 Z_{36} \\
-F_{14} &= -I_1 Z_{14} + I_4(Z_{14} + Z_4 + Z_{45}) - I_5 Z_{45} \qquad (7.1) \\
0 &= -I_2 Z_{25} - I_4 Z_{45} + I_5(Z_{25} + Z_{45} + Z_5 + Z_{56}) \\
&\quad - I_6 Z_{56} \\
F_6 &= -I_3 Z_{36} - I_5 Z_{56} + I_6 (Z_{36} + Z_{56} + Z_6).
\end{aligned}
$$

Each unknown I in these equations can be calculated as a ratio of determinants.

This form of writing the currents leads immediately to one of the well-known theorems that apply to linear circuits: *Superposition Theorem. The current produced anywhere in a linear circuit by the simultaneous action of several voltage generators is the sum of the currents that would be produced there by the generators acting separately.* In our example, the theorem follows directly from the structure of the determinants in the numerator expanded by minors of the voltages F_i, F_{ij}.

Although we shall not do so explicitly, one can also prove another useful theorem from the circuit Equations 7.1. *Thévenin's Theorem. Every linear circuit with two terminals, containing voltage generators all at a common frequency, is, so far as external connections are concerned, equivalent to a voltage generator in series with an impedance. The voltage of the generator is the open-circuit voltage between the terminals, and the impedance is that of the circuit as measured from the terminals, with all generators shorted out.* One can approach this theorem with the aid of an impedance connected between the terminals on loop 3, and a new loop current I_7 passing through this impedance and the Z_0 that is already present.

Thévenin's theorem provides a rationale for rather general application of the result $Z = Z_i^*$ given at the end of Chapter 3 as the condition for maximum power transfer from a system with internal impedance Z_i in series with a voltage source. This result can be applied to complicated circuits equipped with output terminals, for everything inside the circuit is equivalent to a voltage source in series with an impedance Z_i.

Suppose now that we do not wish to know the currents, but rather an output voltage—here taken to be the voltage across the impedance Z_0 in loop 3, as indicated by the terminals drawn in the diagram. This voltage obviously is I_3Z_0, and its calculation requires the determination of only one of the unknown Is. Suppose further that the only generator is F_1, which we shall call the "input" voltage. Then we can write the output voltage I_3Z_0—let us call it E—as a ratio of complex functions of frequency, multiplied by F_1 (now called F):

$$E = FN(\omega)/D(\omega) = h(\omega)\, F, \tag{7.2}$$

where $D(\omega)$ is the determinant of the entire set of impedances, and $N(\omega)$ is Z_0 times the minor of F_1 in the determinant in the numerator of I_3—i.e., Z_0 times the 5×5 determinant that remains when the first row and the third column of the Z determinant are deleted.

Now it is appropriate to discuss one more general circuit theorem: *Reciprocity Theorem. The current produced in loop i by a given voltage source in loop j equals the current produced in loop j by the same voltage source placed in loop i.* In the present case we have one voltage source on loop 1, and are calculating the current in loop 3. The determinant in the numerator is $F_1 \times$ the minor of the 13 element of the original determinant. The reciprocal problem, with F_1 in loop 3 and the unknown current in loop 1, would have in the numerator $F_1\, x$ the minor of the 31 element. This is the same as the other numerator, because the array of impedances in the Equations 7.1 is symmetric. This property of symmetry obviously depends on the signs used in the several equations; in particular, the Equations 7.1 are written with every emf on the left having the sign that makes it aid rather than oppose the current in that loop. The symmetry of the coefficients requires that all emfs be chosen in this way or all in the opposite way. Also, the theorem of reciprocity must be interpreted in terms of emfs that are both with, or both against, the respective currents.

Each of the functions $N(\omega)$ and $D(\omega)$ consists of a sum of products of impedances, each term of which contains (in this case of six loops) six impedances multiplied together. In the general case of n loops, there are n impedances multiplied together in each function. Thus the powers of ω in each term of each function N and D can range from $-n$ to n, in an n-loop network. One can then multiply N and D each by ω^n without changing their ratio $h(\omega)$; now $h(\omega)$ is represented by a ratio of polynomials in ω, each having degree $2n$ or less.

This function $h(\omega)$, which we shall call the system *response function,* plays the same part in the relationship between E and F at a given frequency as is played in the single-loop circuit by the admittance $Y = 1/Z$ in the relationship between I and V. In the single-loop circuit the precise analog of $h(\omega)$ would be the ratio of the voltage across part of the circuit (say across R) to that across the whole circuit. This h, defined for the impedance Z_0 being R, is then

$$\begin{aligned} h(\omega) &= R/Z = R/(i\omega L + R + 1/i\omega C) \\ &= \omega R/(i\omega^2 L + R\omega + 1/iC) \\ &= \omega R/iL(\omega - \omega_1)(\omega - \omega_2), \end{aligned} \tag{7.3}$$

where ω_1 and ω_2 are defined in Equation 6.2. So in this simple case, if F is the complex amplitude of the voltage applied across the whole circuit (input), and the output E is the amplitude of the voltage across R, this $h(\omega)$ relates them:

$$E = h(\omega)F = [\omega R/iL(\omega - \omega_1)(\omega - \omega_2)]F. \tag{7.4}$$

Now, in accordance with our previous work, we can readily deal with nonsinusoidal input and output with the help of Fourier analysis. In particular, if $F(t)$ has a Fourier transform $\alpha(\omega)$ and $E(t)$ is assumed to have a transform $\beta(\omega)$, these transforms are related in the same way as the amplitudes E and F at a given value of ω:

$$\beta(\omega) = h(\omega)\,\alpha(\omega). \tag{7.5}$$

We have verified this relationship explicitly in the mechanical case for the relationship of $x(t)$ to $f(t)$; it is not hard to translate it to the present case and more general cases. In particular, the same relationship 7.5 holds between the transforms of the output voltage $E(t)$ and the input voltage $F(t)$ in a complicated circuit like the example depicted:

If

$$F(t) = \frac{1}{\sqrt{2\pi}} \int_{-\infty}^{\infty} \alpha(\omega) e^{i\omega t}\, d\omega, \tag{7.6}$$

and

$$E(t) = \frac{1}{\sqrt{2\pi}} \int_{-\infty}^{\infty} \beta(\omega) e^{i\omega t}\, d\omega,$$

then

$$\beta(\omega) = h(\omega)\, \alpha(\omega), \tag{7.5'}$$

so

$$E\ (t) = \frac{1}{\sqrt{2\pi}} \int_{-\infty}^{\infty} h(\omega) \alpha(\omega) e^{i\omega t}\, d\omega. \tag{7.7}$$

In our work on a simpler system, we then proceeded to substitute in Equation 7.7 the expression for $\alpha(\omega)$ as the (inverse) transform of $F(t)$; we then reversed the order of the two integrations, and arrived at the Green's-function expression for $x(t)$ (cf. Equations 6.29 and 6.25′). Presumably we can do the same thing here:

$$E(t) = \frac{1}{2\pi} \int_{-\infty}^{\infty} h(\omega) e^{i\omega t} \left[\int_{-\infty}^{\infty} F(t')\, e^{-i\omega t'}\, dt' \right] d\omega; \tag{7.8}$$

$$E\ (t) = \int_{-\infty}^{\infty} G(t - t')\, F(t') dt', \tag{7.9}$$

where

$$G(t - t') = \frac{1}{2\pi} \int_{-\infty}^{\infty} h(\omega) e^{i\omega(t-t')} d\omega \ . \tag{7.10}$$

It is clear on physical grounds that G must be zero when $t < t'$—as we have recognized previously. As in the earlier, simpler case, this requirement implies that the poles of $h(\omega)$ are all in the upper half-plane—h being an algebraic function and thus unable to alter the effect at $|\omega| = \infty$ of the exponential factor in the integrand. So a partial check on a solution of the circuit is the location of these poles. Note that a circuit containing no resistances, only L and C, will have a denominator of h, $D(\omega)$, containing only even powers of ω, that is, D is a polynomial in ω^2. Thus every zero of D at some value of ω has a counterpart at $-\omega$. Since no poles can be in the lower half-plane, these symmetrically located poles must all be on the real axis in this case. Although there are prescriptions for evaluating integrals on paths passing through poles (involving the so-called Cauchy "principal value" of the integral), which in effect amount to counting half the effect of the pole on each side of the contour—in this case such an interpretation is inappropriate. In the analysis of a nondissipative circuit, each pole on the real axis should be interpreted as being *above* the real axis—or the path of integration should be displaced slightly below the axis.

In order to illustrate the calculation of Green's functions by this method, we shall work out two examples. The first case will be the circuit depicted in Figure 7.2. First we solve the case in which $F(t) = Fe^{i\omega t}$ and therefore $E(t) = Ee^{i\omega t}$; we shall define the two loop currents to be positive when clockwise:

$$\begin{aligned} F &= I_1(R + 1/i\omega C) - I_2/i\omega C; \\ 0 &= -I_1/i\omega C + I_2(R + 2/i\omega C). \end{aligned} \tag{7.11}$$

The output voltage E is $I_2/i\omega C$:

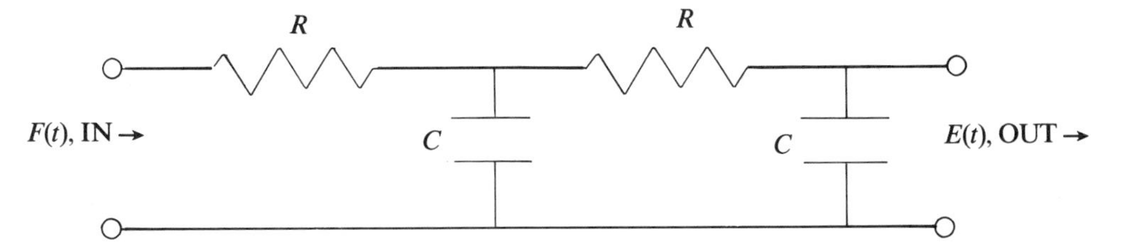

Figure 7.2. A four-terminal RC network.

$$E = \frac{1}{i\omega C} \frac{\begin{vmatrix} R + \frac{1}{i\omega C} & F \\ \frac{-1}{i\omega C} & 0 \end{vmatrix}}{\begin{vmatrix} R + \frac{1}{i\omega C} - \frac{1}{i\omega C} & \\ -\frac{1}{i\omega C} & R + \frac{2}{i\omega C} \end{vmatrix}} = \frac{-F/\omega^2 C^2}{R^2 + \frac{3R}{i\omega C} - \frac{1}{\omega^2 C^2}}$$

$$\equiv h(\omega) F,$$

whence

$$h(\omega) = \frac{-1/R^2 C^2}{\omega^2 - \frac{3i}{CR}\omega - \frac{1}{R^2 C^2}} = \frac{-1/R^2 C^2}{(\omega - \omega_1)(\omega - \omega_2)}, \tag{7.13}$$

where ω_1 and ω_2 are the roots of the denominator,

$$\begin{aligned} \omega_1 &= i(3 + \sqrt{5})/2RC \equiv ib_1 \\ \omega_2 &= i(3 - \sqrt{5})/2RC \equiv ib_2. \end{aligned} \tag{7.14}$$

Reassuringly, these roots are both above the real axis. Thus, applying Equation 7.10, we have $G(t - t') = 0$ when $t < t'$. When $t > t'$, we integrate around C_+, getting

$$G(t - t') = 2\pi i\,(1/2\pi)\,(-i/R^2C^2)\left[\frac{e^{i\omega_1(t-t')}}{\omega_1 - \omega_2} + \frac{e^{i\omega_2(t-t')}}{\omega_2 - \omega_1}\right]$$

$$= \frac{-1}{R^2C^2(b_1 - b_2)}\,[e^{-b_1(t-t')} - e^{-b_2(t-t')}] \tag{7.15}$$

$$= \frac{1}{\sqrt{5}\,RC}\,[e^{-[(3-\sqrt{5})/2RC](t-t')} - e^{-[(3+\sqrt{5})/2RC](t-t')}].$$

Note, the requirement of poles' being in the upper half-plane has not only guaranteed "causality" (effects not preceding causes) but also it has guaranteed that the singular values of ω, substituted into $e^{i\omega(t-t')}$ in the calculation of residues, produce exponential functions in G that *decrease* with time. An increasing exponential would be almost as unreasonable physically as would an output that preceded the input.

We should keep in mind, also, that the Green's function not only mediates between a general input and a general output, as in Equation 7.9; also it is the response of the system to a delta-function input. One's experience with circuits may impart enough feeling for their behavior to permit a judgment of the plausibility of a given Green's function as the output arising from a delta-function input at time t'. It is recommended that all results of calculations like those being explained here should be checked for plausibility by all available means; not only can mistakes be thus detected but insight can be trained in this way, too.

Our second example will illustrate a difficulty that can interfere with the calculation of Green's functions and will show how to deal with the problem. In Figure 7.3, let us simply interchange the resistors with the capacitors in Figure 7.2. As before, we calculate $h(\omega)$ by assuming a single frequency ω fed into the circuit. $E = RI_2$, and we determine I_2 from

$$F = I_1(R + 1/i\omega C) - I_2R;$$
$$0 = -I_1R + I_2(2R + 1/i\omega C). \tag{7.16}$$

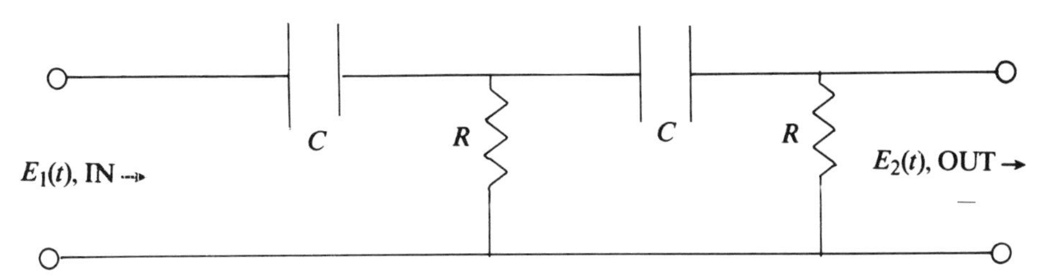

Figure 7.3. The network of Figure 7.2, with R and C interchanged.

Thus

$$E = h(\omega)\,F = R\,\frac{\begin{vmatrix} R + \dfrac{1}{i\omega C} & F \\ -R & 0 \end{vmatrix}}{\begin{vmatrix} R + \dfrac{1}{i\omega C} & -R \\ -R & 2R + \dfrac{1}{i\omega C} \end{vmatrix}} = \frac{FR^2}{R^2 + \dfrac{3R}{i\omega C} - \dfrac{1}{\omega^2 C^2}}, \tag{7.17}$$

so

$$h(\omega) = \frac{\omega^2}{\omega^2 - \dfrac{3i\omega}{RC} - \dfrac{1}{R^2C^2}} = \frac{\omega^2}{(\omega - \omega_1)(\omega - \omega_2)}, \tag{7.18}$$

where the roots of the denominator are the same quantities that are in Equation 7.14. To this extent, the present example is like the preceding one. But this $h(\omega)$, unlike the other one, fails to go to zero as $|\omega|$ goes to infinity. Thus, according to Equations 6.4 to 6.9, the integrals over the curved paths S_+ and S_- do not vanish even though the exponential factor in the integrand of Equation 7.10 may properly go to zero. In fact, substituting this h into Equation 7.10, we see that the integral seems to diverge regardless of contours and residues:

$$G(t - t') = \frac{1}{2\pi}\int_{-\infty}^{\infty} \frac{\omega^2 e^{i\omega(t-t')}\, d\omega}{(\omega - \omega_1)(\omega - \omega_2)}. \tag{7.19}$$

Such an integral is reminiscent of the similarly divergent integral for the delta function:

$$\delta(t - t') = \frac{1}{2\pi}\int_{-\infty}^{\infty} e^{i\omega(t-t')}\, d\omega. \tag{7.20}$$

In fact by subtracting this divergent integral from the one for G, we get an integral that converges:

$$\begin{aligned} G_s(t - t') &\equiv G(t - t') - \delta(t - t') \\ &= \frac{1}{2\pi}\int_{-\infty}^{\infty}\left[\frac{\omega^2}{(\omega - \omega_1)(\omega - \omega_2)} - 1\right] e^{i\omega(t-t')} d\omega \\ &= \frac{1}{2\pi}\int_{-\infty}^{\infty}\frac{3i\omega/RC + 1/R^2C^2}{(\omega - \omega_1)(\omega - \omega_2)} e^{i\omega(t-t')} d\omega \\ &\equiv \frac{1}{2\pi}\int_{-\infty}^{\infty} h_s(\omega) e^{i\omega(t-t')}\, d\omega, \end{aligned} \tag{7.21}$$

where the subscript s stands for "subtracted," and $h_s(\omega) = h(\omega) - 1$. Now the output voltage $E(t)$ is given by

$$E(t) = \int_{-\infty}^{\infty} [G_s(t - t') + \delta(t - t')] F(t')\, dt' = F(t) + \int_{-\infty}^{\infty} G_s(t - t')\, F(t')\, dt'. \tag{7.22}$$

We shall not evaluate the subtracted Green's function G_s for the foregoing example, because it now has no new features that have not already been seen in other examples. What is noteworthy here is that, when $h(\omega)$ approaches a constant as $|\omega| \to \infty$, the output equals the input *plus* a typical Green's function integral; the input voltage appears at the output with no time delay and is then followed by a delayed response.

Other circuits in which $h(\omega)$ produces divergence of the integral for G are those in which all circuit elements are of the same type, all R or all L or all C. Then $h(\omega)$ is independent of ω, so G is a multiple of the delta function without a G_s. In particular, a purely resistive network instantaneously transmits the input (attenuated) to the output.

One of the predictions of the special theory of relativity is that no signal can travel faster than c, the speed of light in vacuo. The foregoing discussion of causality and Green's functions—not only the delta-function term that is in some of them but also the Green's functions that rise slowly—predicts that the output will begin at the instant when the input begins. This prediction, applied to a circuit of nonzero length, contradicts the theory of relativity. The contradiction occurs because the properties that we have assumed for electric circuits are only approximately correct. The assumption of linearity and freedom from hysteresis are accurate enough, but it is not justified to assume that circuits can be made with certain specified elements at certain places and no others anywhere. Any circuit has small capacitances between conductors, and every conductor has some inductance. Furthermore, circuits carrying high-frequency currents emit some electromagnetic radiation which takes away some energy and thus, by leaving the circuit, causes some resistance to be present that would not be there in the absence of radiation. These effects, if accurately taken into account, lead to the finite limiting speed of signals that is predicted by relativity, but we shall not try to include them. We shall simply have to remember that instantaneous action at a distance is really not quite instantaneous, after all. In a circuit with 10 cm between input and output, for example, there must be a time delay of 3.3×10^{-10} sec, corresponding to a frequency of about 3×10^9 Hz; such a time delay can be ignored in a circuit in which the typical frequency in use is no higher than about 10^8 Hz.

The discussion so far has concerned systems that, although they were subjected to specified forces or emfs, were isolated in that all the impedances had been explicitly taken into account. Let us now briefly consider what can be said about a system with two input and two output terminals when these terminals are connected to unknown external systems. We shall assume that the system is passive (i.e., has no internal voltage sources) and that it contains only linear circuit elements and only one frequency, ω. Figure 7.4 partly depicts such a system, with loops 1 (input) and n (output) shown, and $n-2$ internal loops not shown. Now, even though we do not know all their causes, we can still make use of the potential differences (complex, of frequency ω) across the input and across the output. Let us call these V_1 and V_n. Correspondingly, we shall let I_1 and I_n be the complex loop currents in the input and the output loops. Each potential difference equals the appropriate sum of $I \times Z$ products; as our circuit is drawn,

$$\begin{aligned} V_1 &= (I_1 - I_2)\, Z_{12} \\ 0 &= \ldots \\ 0 &= \ldots \\ V_n &= (I_{n-1} - I_n) Z_{n-1,n} \end{aligned} \quad \left.\begin{matrix} \\ \\ \end{matrix}\right\} \; (n - 2 \text{ equations}). \tag{7.23}$$

There are n equations in a set – enough, as we have seen, to determine all the I_k if V_1 and V_n are known. But it is *not* enough to know only the input voltage V_1, for I_n does not entirely determine V_n. However, one can solve for everything else by knowing V_1 *and* I_1. Instead of discussing a direct solution based on such information, let us proceed indirectly by assuming first that the currents have all been expressed in terms of V_1 and V_n, in the conventional manner; let us consider only the solutions for I_1 and I_n,

Figure 7.4. The input and output of a network in which V_1 and I_1 determine V_n and I_n, regardless of external connections.

which will involve ratios of determinants that can be abbreviated as

$$\begin{aligned} I_1 &= y_{11}V_1 + y_{1n}V_n; \\ I_n &= y_{n1}V_1 + y_{nn}V_n. \end{aligned} \tag{7.24}$$

Now, as previously argued, the array of coefficients in the circuit equations can always be made symmetric, as in Eq.(7.1), by appropriate choice of the signs of the emfs on the left. In this case we could argue that $y_{1n} = y_{n1}$ in (7.24). However, we have chosen V_1 to aid I_1 and V_n to oppose I_n; i.e., if we reversed the sign of one of the two V's we would have a symmetric array. As they are, though, $y_{1n} = -y_{n1}$. Thus, solving (7.24) for V_n and I_n in terms of V_1 and I_1, we have

$$\begin{aligned} V_n &= -(y_{11}/y_{1n})V_1 + I_1/y_{1n} = AV_1 + BI_1; \\ I_n &= -(y_{1n} + y_{nn}y_{11}/y_{1n})V_1 + (y_{nn}/y_{1n})I_1 = aV_1 + bI_1. \end{aligned} \tag{7.25}$$

We can readily verify that the determinant of the coefficients equals 1.

The coefficients in Equation 7.25 are functions of frequency. We can reason somewhat as we have done before, to apply Fourier analysis to the determination of V_n and I_n when V_1 and I_1 are nonsinusoidal. Each of the output quantities will be expressed as a sum of Fourier series or integrals, one arising from V_1 and the other from I_1. Similarly, Green's functions can be derived for the network that has given frequency-dependent A, B, a, b. There will be four Green's functions.

NOTES

References on the theory of electrical networks are:

Johnson, D. E. *Introduction to Filter Theory,* 1976, Prentice Hall, Englewood Cliffs, NJ

Christian, E., *LC Filters: Design, Testing, and Manufacture,* 1983, John Wiley and Sons, New York

Van Valkenburg, M. E. *Network Analysis.* 3d ed. Englewood Cliffs, NJ: Prentice-Hall, 1974.

PROBLEMS

7.1. In the depicted circuit $R > \frac{1}{2}(L/C)^{\frac{1}{2}}$. Calculate a Green's function so that $E_2(t)$ can be calculated in terms of $E_1(t)$.

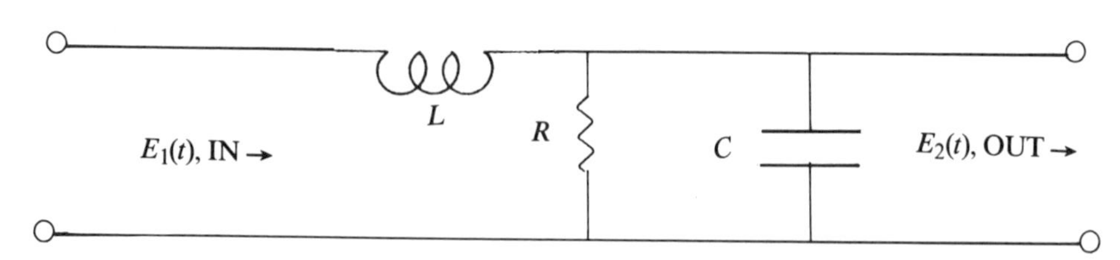

7.2. Calculate a similar Green's function for this circuit.

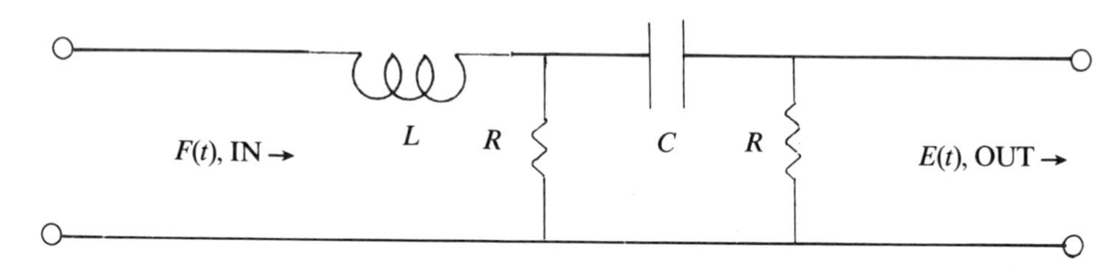

7.3. And for this one:

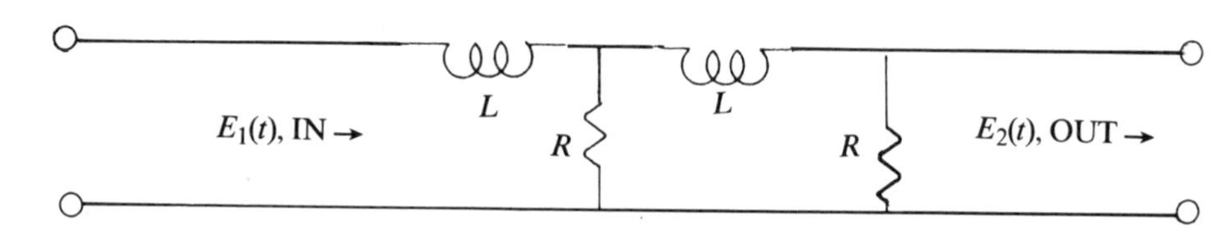

7.4. And this:

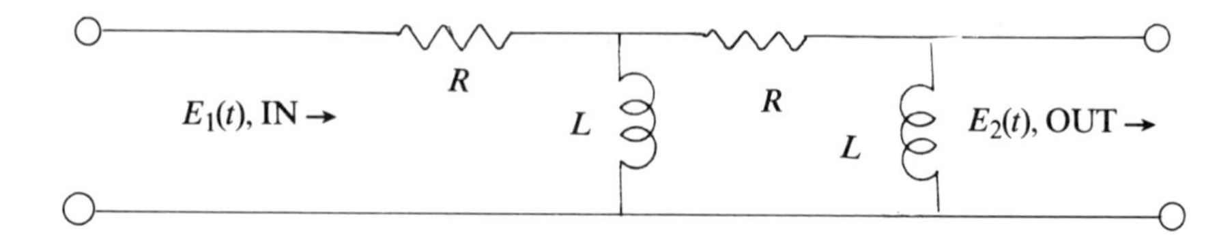

7.5. In terms of inductances, resistances, and capacitances, what conditions must the impedances in this circuit satisfy to make the integral for the Green's function converge without subtraction?

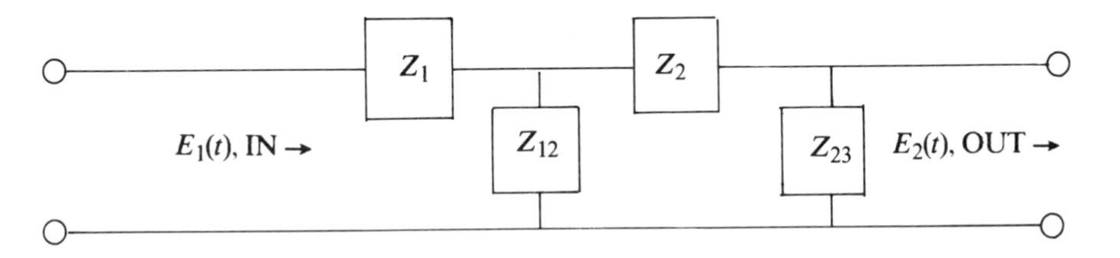

8

Nonlinear Oscillations

The main purpose of this book is to acquaint students with some physical systems describable by linear mathematics and to introduce them to the elegant and systematic mathematical methods available for understanding such systems. This is an important purpose, because some of the systems studied by physicists are linear, or approximately so, and because a thorough knowledge of linear systems can improve one's insight into nonlinear systems. There has been much work on nonlinear problems in recent years, work that physicists should study, because many of the systems that they wish to understand are either slightly or seriously nonlinear. Therefore we devote this chapter and one other to introductory discussion of a few simple nonlinear systems.

We pointed out in the first two chapters that mechanical, and some electrical, oscillators behave linearly only in the limit of small oscillations. A single undamped oscillator such as a pendulum or a bead sliding on a wire has a restoring force that contains higher powers of x; their equation of motion has the form

$$m\ddot{x} = f(x). \tag{8.1}$$

This equation can be integrated once immediately; we multiply by $\dot{x}$:

$$m\ddot{x}\dot{x} = \tfrac{1}{2}m d\dot{x}^2/dt = f(x)\dot{x} \equiv -dU(x)/dt, \tag{8.2}$$

where

$$f(x) = -dU/dx. \tag{8.3}$$

Thus we obtain the familiar result that

$$T + U = \tfrac{1}{2}m\dot{x}^2 + U(x) = E = \text{constant}. \tag{8.4}$$

This equation can be integrated immediately:

$$\int \frac{dx}{\sqrt{E - U}} = \sqrt{\frac{2}{m}}\,(t - t_0). \tag{8.5}$$

This step solves the problem, but the integral may be hard to evaluate, and if one does get $t(x)$ explicitly, one may still have trouble getting $x(t)$, which is usually what is required.

This solution was possible because the system is conservative: energy is a constant of the motion. It is well known, also, that more complicated systems can be partly integrated by use of constants of the motion. For example, the problem of two particles in three dimensions that are subject only to the forces that they exert on each other can be fully solved if the forces are conservative and central. Then there are seven constants of the motion which allow the problem to be reduced to quadratures, the three components of total momentum, the three components of angular momentum of relative motion, and the energy of relative motion. The center of mass moves with constant velocity, and four of the six integrations needed to give relative motion of the two masses are done immediately with the help of conservation laws. The result is that the relative motion is plane and, if r, θ are relative polar coordinates in that plane and μ is reduced mass,

$$\tfrac{1}{2}\mu(\dot{r}^2 + r^2\dot{\theta}^2) = E - U(r); \tag{8.6}$$

$$\mu^2 r^4 \dot{\theta}^2 = l^2. \tag{8.7}$$

Here E and U are, respectively, total and potential energy, and l is angular momentum, associated with the particles' relative position and velocity. We substitute from Equation 8.7 for the $\dot{\theta}^2$ in Equation 8.6; we get

$$\tfrac{1}{2}\mu\dot{r}^2 + U(r) + \frac{l^2}{2\mu r^2} \equiv \tfrac{1}{2}\mu\dot{r}^2 + V(r) = E, \tag{8.8}$$

where

$$V(r) = U(r) + \frac{l^2}{2\mu r^2} \tag{8.9}$$

is the effective potential energy affecting radial motion. The l^2 term, known as the centrifugal potential, is really kinetic energy of azimuthal motion. Equation 8.8 can be integrated immediately:

$$t(r) - t_0 = \int \frac{dr}{\sqrt{\frac{2}{\mu}\,[E - V(r)]}}. \tag{8.10}$$

This equation is analogous to Equation 8.5, and is subject to the same disadvantages: the integral may be hard to do and an expression for $t(r)$ may be hard to solve for $r(t)$.

It is possible also to eliminate t from Equations 8.6 and 8.7 by dividing the former by the latter:

$$\frac{1}{2\mu r^4}\left[\left(\frac{dr}{d\theta}\right)^2 + r^2\right] = \frac{E - U(r)}{l^2}, \tag{8.11}$$

whence

$$\theta(r) - \theta_0 = \int \frac{l\,dr}{r\sqrt{2\mu r^2[E - U(r)] - l^2}}$$

$$= \frac{l}{\sqrt{2\mu}} \int \frac{dr}{r^2\sqrt{E - V(r)}} \,. \tag{8.12}$$

Equation 8.12, when integrated, describes the orbit of the system in the plane of relative motion; normally one prefers to know $r(\theta)$ rather than $\theta(r)$.

The foregoing systems have several features that are worth summarizing:

1. The systems are capable of oscillatory motion, provided the potential U of the one-dimensional system or the effective potential V of the two-body system has a relative minimum. If the system is located between two maxima of V with energy less than both these maximum values, it will move periodically—sinusoidally if the energy is only slightly above the minimum. In the central-force case this periodic motion is radial oscillation around a circular orbit. If the energy is sufficient to permit r or x to go to infinity, the motion is not periodic.
2. Except in the limit of small oscillations, the systems have nonlinear equations of motion, which lack the convenient property that linear combinations of solutions are also solutions. In particular, oscillatory solutions cannot be assigned arbitrary amplitudes; their amplitudes are likely to depend on their frequency spectra.
3. The systems are conservative; they do not dissipate energy. This property is related to, but not the same as, reversibility of motion. The latter property, sometimes called time-reversibility, implies that every possible motion is converted to a possible motion by the reversal of the sign of t, that is, by the transformation $t \rightarrow -t$. This behavior of the solutions of the equations of motion is equivalent to the invariance of these equations under the time-reversal transformation. Equations of motion containing no functions of t and only even time derivatives and even powers of odd derivatives have this invariance. Equations that do not have it are those in Chapter 2 that describe damped harmonic motion, for these equations contain first time derivatives which reverse sign under time reversal. No dissipative system can have reversible motion,

for obvious reasons. There are also nondissipative systems whose motion is irreversible, e.g., a charged particle moving in a magnetic field. Unlike dissipative forces, such magnetic forces do no work (being perpendicular to velocity) and are thus derivable from a (velocity-dependent) potential.

In the next chapter a systematic procedure is discussed for dealing with conservative systems, including some in which there are forces with explicit time dependence. Such problems are seldom easy, but many approaches to them have been known for a long time; they are discussed at length in books on classical mechanics.

For the present, we shall treat systems whose equations of motion contain first time derivatives. If such derivatives appear only raised to even powers they still permit time-reversibility. We shall suppose that first derivatives may enter the equations in any way, but shall simplify our discussion by dealing with a system that has only one coordinate, x, with an equation of motion

$$\ddot{x} = Q(x, \dot{x}). \tag{8.13}$$

It is often useful to define $y = \dot{x}$, and work in the xy-plane, which is called the *phase plane*. Strictly speaking, *phase space*—or in this case the phase plane—has axes corresponding to all the coordinates and all the momentum components of a system, instead of coordinates and velocity components. However, the two ways of labeling axes are equivalent whenever, as in our case, each momentum component is a constant multiple of a velocity component; our usage is thus justified. The advantage of describing systems' behavior in phase space is that a point in that space uniquely determines the mechanical *state* of a system, i.e., it determines a complete set of initial conditions for the equations of motion and thus determines the future motion of the system. So the second-order Equation 8.13 becomes two first-order equations:

$$\begin{aligned} \dot{y} &= Q(x, y), \\ \dot{x} &= y, \end{aligned} \tag{8.14}$$

which is a special case of

$$\begin{aligned} \dot{y} &= Q(x, y), \\ \dot{x} &= P(x, y). \end{aligned} \tag{8.15}$$

As in the case of Equations 8.6 and 8.7, one can eliminate t from these equations by dividing one by the other:

$$\frac{dy}{dx} = \frac{Q(x, y)}{P(x, y)}, \tag{8.16}$$

or

$$Pdy - Qdx = 0. \tag{8.17}$$

If the differential in this equation is exact, that is, if $Pdy - Qdx$ can be written as $dH(x, y)$, the equation states that

$$H(x, y) = \text{constant} \tag{8.18}$$

The necessary and sufficient condition for the differential $Pdy - Qdx$ to be exact is that

$$\partial P/\partial x = -\partial Q/\partial y. \tag{8.19}$$

A case in which this condition is satisfied is that described in Equation 8.1, in which $Q(x, y) = f(x)$ and $P(x, y) = y$. Condition 8.19 is obviously satisfied, and

$$F(x, y) = \int (Pdy - Qdx) = \tfrac{1}{2}y^2 - \int f(x)dx = \text{const.}, \tag{8.20}$$

which is equivalent to Equation 8.4. If $Pdy - Qdx = dH(x, y)$, Equations 8.15 imply that

$$\begin{aligned} \dot{y} &= Q(x, y) = -\partial H/\partial x; \\ \dot{x} &= P(x, y) = \partial H/\partial y. \end{aligned} \tag{8.21}$$

This is the so-called *Hamiltonian* form of the equations of motion, and the function $H(x, y)$ is called the *Hamiltonian* function (properly so-called only when y is the momentum, rather than the velocity).

But if the force $f(x)$ depends also on $\dot{x} = y$, Equation 8.19 is no longer satisfied. The damped linear system of Chapter 2 has $P = y, Q = -2\gamma y - \omega_0^2 x$, and thus violates Equation 8.19. There is, however, a theorem concerning equations of the form of Equations 8.17 to the effect that such an equation in two variables always has an integrating factor $g(x, y)$, that is, there exists a function g such that $gPdy - gQdx$ is exact. P and Q being given, substituting gP and gQ respectively for P and Q in Equation 8.19 produces a partial differential equation for the integrating factor g. Usually it is harder to solve this equation for an acceptable real g and then to use g in a line integral of $gPdy - gQdx$ in the xy-plane than it is to attack Equation 8.16 by a different method. For example, sometimes the variables can be separated, as in the trivial case $P = P(x)$, $Q = Q(y)$; one separates variables by inspection instead of formally solving for $g = 1/PQ$.

Another approach that sometimes permits analysis of the problem is to change variables in the xy-plane to some others such as polar coordinates. In order to do this, we first adopt a dimensionless time variable. In the familiar case of the damped linear system, we can define

$$\tau = \omega_0 t, \tag{8.22}$$

so that, using primes to denote τ derivatives, we get the equation of motion

$$x'' + 2sx' + x = 0, \tag{8.23}$$

where

$$s = \gamma/\omega_0. \tag{8.24}$$

Letting $y = x'$, we have

$$y' = Q(x, y) = -x - 2sy,$$

$$x' = P(x, y) = y,$$

so

$$\frac{dy}{dx} = \frac{Q}{P} = \frac{x}{y} - 2s. \tag{8.25}$$

Now, changing to polar coordinates r, θ in the xy-plane, we have (after some rearranging)

$$\frac{dr}{rd\theta} = \frac{\cos\theta + (Q/P)\sin\theta}{(Q/P)\cos\theta - \sin\theta} = \frac{s(1 - \cos 2\theta)}{1 + s\sin 2\theta}. \tag{8.26}$$

Here the middle member of the equation is the general expression for the logarithmic derivative on the left, and the right-hand member is the form that this expression takes in the case under discussion.

It is not particularly hard to integrate this equation for the linear case, for it is separated as it stands and the θ integral is an easy one. However, we have to be cautious in inserting limits such as $0, 2\pi$ to get the change in $\log r$ in one cycle of oscillation; the θ integral is a multiple-valued function, and we must be sure to insert the correct value at each limit. The expression on the right is clearly nonnegative if $s < 1$, so $\log r$ is an increasing function of θ. The definition of y implies that x is increasing with time when y is positive and decreasing when y is negative. Thus with the passage of time the point representing the state of the system is encircling the origin clockwise; i.e., θ decreases as x (or t) increases. Thus $\log r$ is decreasing with the passage of time, and asymptotically approaches the origin, which is the only point where x' and y' both vanish—thus the only equilibrium point for the system. Such a state of a system, approached from all states in a finite area of the phase plane, has come to be called an *attractor.* The path of the system in the phase plane is a spiral which winds inward toward the origin as one follows it clockwise. (See Figure 8.1.) This spiral shows the sequence of states through which the system passes, but conveys no information about the rate at which it traverses its path.

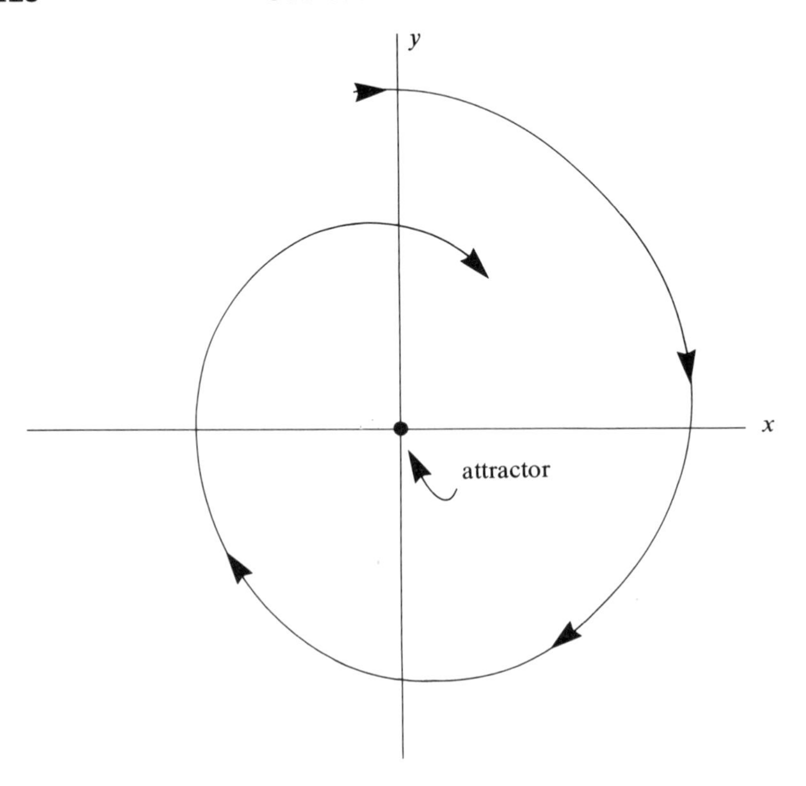

Figure 8.1 The trajectory of the linear, undriven damped oscillator in the phase plane

The damped, undriven linear oscillator is thoroughly understood, so does not require the kind of analysis outlined here. Because of its familiarity, however, it is useful for illustrating methods that may be the only ones available for analysis of some nonlinear systems. Even with the best available methods, some systems may be amenable only to qualitative analysis, methods of successive approximation, or numerical integration. We shall next look at two rather artificial examples, contrived so as to be integrable, and then shall briefly discuss the use of the phase plane for qualitative analysis of nonintegrable systems. We alter Equation 8.23 by allowing the damping term to become nonlinear:

$$x'' + fx' + x = 0, \tag{8.27}$$

where f is an unspecified function of the coordinates in the phase plane. Then Equations 8.25 become

$$y' = -x - yf, \qquad x' = y. \tag{8.28}$$

On changing to polar coordinates, we have

$$r' = -rf\sin^2\theta = -\tfrac{1}{2}rf(1 - \cos 2\theta), \qquad \theta' = -1 - f\sin\theta\cos\theta = -\tfrac{1}{2}(2 + f\sin 2\theta), \tag{8.29}$$

which lead to

$$\frac{dr}{rd\theta} = \frac{f(1 - \cos 2\theta)}{2 + f\sin 2\theta}, \tag{8.30}$$

as in Equation 8.26. Rewriting this as

$$[(2/f + \sin 2\theta)/r]dr = (1 - \cos 2\theta)\, d\theta, \tag{8.31}$$

we apply the condition for integrability (Equation 8.19), with r, θ substituted for x, y:

$$\frac{\partial}{\partial\theta}\frac{1}{f} + 2\cos 2\theta = 0, \tag{8.32}$$

the solution to which is

$$f = \frac{1}{g(r) - \frac{1}{2}\sin 2\theta}, \tag{8.33}$$

where $g(r)$ is an arbitrary function of r. This choice for f renders Equation 8.31 separable:

$$2g(r)\, dr/r = (1 - \cos 2\theta)\, d\theta. \tag{8.34}$$

Thus, integrating, we have

$$2 \int g(r)\, dr/r = \theta - \tfrac{1}{2} \sin 2\theta + C. \tag{8.35}$$

It is clear from Equation 8.29 that r increases with t if $f < 0$ and decreases if $f > 0$, and that a circular orbit in the phase plane occurs if and only if $f = 0$ for all θ. According to Equation 8.33, $f = 0$ is possible only when $g(r) = \pm\infty$. A case in which that happens is

$$g(r) = Ar \tan kr, \tag{8.36}$$

with A and k constant. This g leads to orbits that approach circularity as r approaches $(n + \frac{1}{2})\,\pi/k$. Note, however, that, if $A > 0$, $f > 0$ when r is just inside the circular orbit and $f < 0$ just outside. This implies that a value of r close to one of the values that make f vanish, either inside or outside, moves farther from that value with the passage of time; the circular orbits are unstable. But if $A < 0$ the motions of r are reversed near the circular orbits; the circular orbits are stable. The function $g(r) = Br \cos kr$ produces circular orbits with similar stability or instability if the sign of B is opposite that of A.

One of the stable circular orbits in the phase plane is called a *limit cycle;* it represents the motion approached asymptotically by a system that starts anywhere in a ring-shaped region of the plane. Thus there are, in such cases, infinitely many attractors which are now not points but steady states of motion.

Any one of the limit cycles is characterized by the damping term in Equation 8.27 being and remaining zero. Thus the system moves like a conservative system, in particular, an undamped oscillator—it moves with constant r, that is, with $\frac{1}{2}y^2 + \frac{1}{2}x^2 = E/m\omega_0^2$ constant.

The motion of the system as it approaches a limit cycle is more complicated; it may be characterized by either increasing or decreasing energy according as r is increasing or decreasing toward one of its stable values. In order to have an equation of motion like Equation 8.27 with f as in Equation 8.33, a system has to have a source of energy, for it has negative damping when r is slightly less than one of its stable values. We have not pretended to derive Equations 8.27 and 8.33, and probably no physical system satisfies just these equations. But there are systems that effectively have negative damping in some circumstances, systems like transistor or vacuum-tube oscillators that get energy from DC power supplies and convert it into energy of oscillation. An equation that describes such a system, the van der Pol equation, will be discussed below.

To solve explicitly for the motion with given initial conditions, one first evaluates the integral in Equation 8.35, choosing C to fit the initial condition, and then solves for the function $r(\theta)$, the equation of the orbit in the phase plane. Thus one can express f in terms solely of θ by Equation 8.33, then substitute this expression in the θ' Equation 8.29. Solving the resulting equation gives $\tau(\theta)$, which one has to solve for $\theta(\tau)$. Then this function can be substituted in the function $r(\theta)$ to yield $r(\tau)$. These procedures, though clearly describable, can be difficult. In the particular case of $g(r)$ being a tangent or a cotangent, the integral encountered in the solution of 8.29 is quite intractable.

We have remarked that these particular choices for $g(r)$ are unrealistic. Another way in which they are unrealistic (when $A > 0$ or $B < 0$) is common to all functions $g(r)$ that have positive derivatives where $g(r) = \frac{1}{2}\sin 2\theta$. At such a point f is singular, and on each side of the point the sign of r' moves the system toward the singularity. Equation 8.30 shows that a singularity of f causes no misbehavior of the orbit in the phase plane, but Equations 8.29 show that such a singularity produces singularities in r' and θ'—the state of the system moves with infinite speed in the phase plane. Such behavior results physically from infinite forces, which are clearly impossible. It would appear that realistic functions $g(r)$ must everywhere have absolute values greater than $\frac{1}{2}$, or must have negative slope wherever they are smaller.

The foregoing discussion of a linear oscillator with a nonlinear damping force can be generalized to other oscillatory one-dimensional systems. If $U(x)$ is the potential energy (divided by mass times a constant of dimensions ω^2), it will enter the equations of motion through its derivative U_x:

$$y' = -U_x - yf;$$
$$x' = y. \tag{8.37}$$

Now we replace the polar coordinates by new, similar ones that are adapted to the present $U(x)$:

$$y = R\sin\theta,$$
$$(2U)^{1/2} = R\cos\theta, \tag{8.38}$$

or

$$R^2 = y^2 + 2U,$$
$$\theta = \tan^{-1}[y/(2U)^{1/2}]. \tag{8.39}$$

Note that R^2 is proportional to total energy, as r^2 was before. We have simply replaced x by $(2U)^{1/2}$, a procedure that may lead to trouble unless U is monotonic for each sign of x. It is necessary also to understand that $(2U)^{1/2}$ is to be taken as positive when x is and negative when x is negative.

Now we write

$$RR' = yy' + yU_x = -RU_x\sin\theta - R^2f\sin^2\theta + RU_x\sin\theta$$
$$= -R^2f\sin^2\theta \tag{8.40}$$

$$\theta' = \frac{(2U)^{-1/2}y' - (2U)^{-3/2}y^2U_x}{1 + y^2/2U}$$
$$= \frac{-R\cos\theta(U_x + Rf\sin\theta) - R^2U_x\sin^2\theta/R\cos\theta}{R^2}$$

so

$$R'/R = -f\sin^2\theta$$
$$\theta' = -f\sin\theta\cos\theta - U_x/R\cos\theta \tag{8.41}$$

Thus

$$\frac{dR}{Rd\theta} = \frac{\sin^2\theta}{\sin\theta\cos\theta + U_x/Rf\cos\theta}, \quad (8.42)$$

or

$$(\sin\theta\ \cos\theta + U_x/Rf\cos\theta)dR/R = \sin^2\theta d\theta. \quad (8.43)$$

The integrability condition then implies that

$$R\ \cos 2\theta + \partial/\partial\theta(U_x/f\cos\theta) = 0, \quad (8.44)$$

whence

$$U_x/f\mathrm{R}\cos\ \theta = -\tfrac{1}{2}\ \sin 2\theta + g(R), \quad (8.45)$$

where g is arbitrary, as before. Thus

$$f = \frac{U_x}{(2U)^{1/2}} \times \frac{1}{g(R) - \frac{1}{2}\sin 2\theta}. \quad (8.46)$$

Substituting this expression into Equation 8.43, we get

$$\int g(R)\ dR/R = \int \sin^2\theta\ d\theta = \tfrac{1}{2}\theta - \tfrac{1}{4}\sin 2\theta + C. \quad (8.47)$$

If the coordinates R, θ are to be usable, we must place the origin of x at the minimum of U, choosing the minimum value of U to be zero. In order to carry out the calculation, we have to express $U^{-1/2}U_x$ (which appears in Equation 8.46) in terms of U itself, so it can then be expressed, by equation 8.39, in terms of R, θ. For example, if U = A(cosh kx -1), one gets $U^{-1/2}U_x = k(U + 2A)^{1/2} = k(\frac{1}{2}R^2\cos^2\theta + 2A)^{1/2}$. Such an expression, combined with R(θ) as determined from Equation 8.47, can be used in Equation 8.41 for θ'.

If the function $g(r)$ in Equations 8.33–8.35 produces limit cycles, the same function of R in the present case produces them also. The limit cycles represent the only possible periodic motions of the systems. Such motions can have only certain discrete amplitudes in the present nonlinear problems, in contradistinction to linear problems, in which all amplitudes are possible. In the case of the harmonic oscillator with nonlinear damping, all limit cycles represent sinusoidal motion with the common frequency ω_0, whereas in the generalizations contemplated above the motion is nonsinusoidal and has a fundamental frequency that depends on the amplitude. In all the foregoing cases, each limit cycle represents one of the possible stable motions with $f = 0$, that is, without damping; each has constant energy.

Not all one-dimensional oscillations are susceptible to such easy precise analysis, but others can be understood qualitatively and analyzed by approximate methods. Let us consider, for example, an undriven *RLC* circuit in which the inductance has an iron core that saturates. As shown in Chapters 1 and 2, the equation describing the behavior of this circuit is

$$V_L + RI + \frac{1}{C}\int Idt = 0, \tag{8.48}$$

where we have temporarily expressed the voltage across the inductance as V_L. By Faraday's law, this voltage equals the rate of change of the circuit's flux linkage. Hitherto we have assumed that all the flux linkage occurs in the coil, and that it is proportional to the current in the coil. But, as mentioned in Chapter 1, iron cores have nonlinear relationships between current and magnetization, so iron-core coils have nonlinear relationships between current and flux linkage. Such a relationship is shown in Figure 8.2. Here ϕ is the flux linkage, and I, as before, is current. This relationship can, within limits, be approximated as

$$I = \phi/L_0 + A\phi^3. \tag{8.49}$$

Thus, taking the time derivative of Equation 8.48 and substituting from Equation 8.49, we have

$$\frac{d^2\phi}{dt^2} + R(1/L_0 + 3A\phi^2)\,\frac{d\phi}{dt} + \phi/L_0C + A\phi^3/C = 0. \tag{8.50}$$

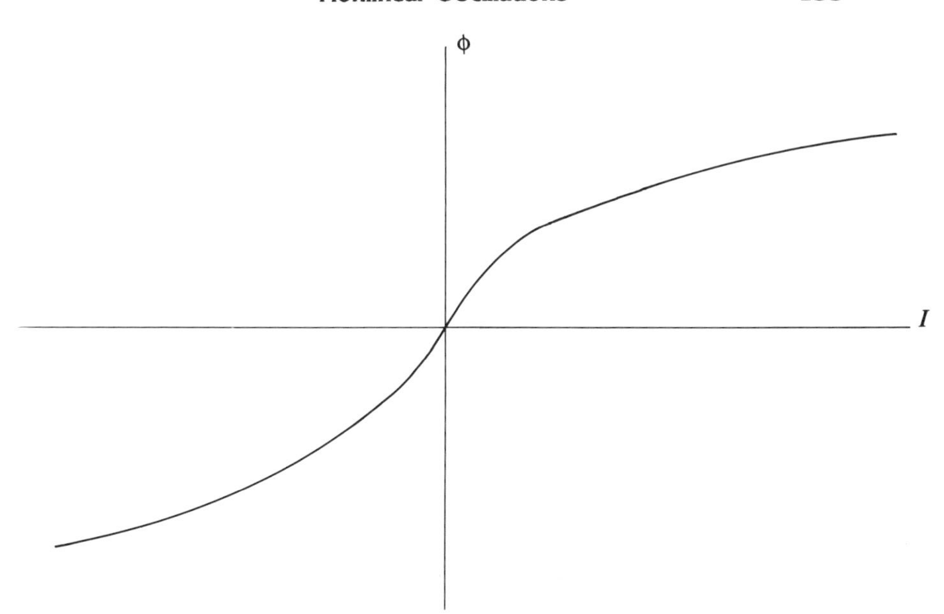

Figure 8.2 A typical relationship of flux linkage ϕ through a coil to current I in the coil. Not shown are the straight portions of the curve at very large values of $|I|$, and the phenomenon of hysteresis.

This equation is only as good as the approximate expression for $I(\phi)$, Equation 8.49, and still neglects hysteresis. Accurate inclusion of hysteresis would introduce into Equation 8.50 terms that take into account not only the value of ϕ at one time but also past values, for Equation 8.49 would contain a constant term and coefficients depending on the history of the system. The overall effect of hysteresis is to produce additional dissipation of energy; it adds to the effective resistance in a series circuit.

It is straightforward to choose a new time variable, $t(L_0C)^{-1/2}$, define $\phi = x$ and $\phi' = y$, and study the orbits in the phase plane. It is clear on physical grounds, or by inspection of Equation 8.50, that the system has no limit cycles, for dissipation is always positive; there are no energy sources that can maintain or stabilize limit cycles. Thus the motion is not strikingly different from that discussed in Chapter 2—the oscillations are damped toward zero. Because of the cubic terms, the motion is not sinusoidal and the methods that employ complex functions and depend on superposition are no longer applicable.

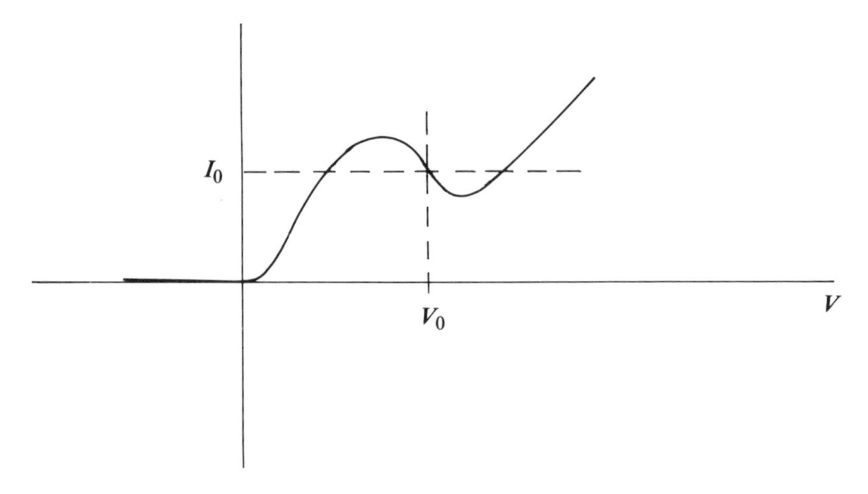

Figure 8.3 A current-voltage curve with a negative-resistance (negative-slope) portion. Note that the ratio V/I is everywhere positive or zero.

Some self-sustaining oscillations are caused by devices that have negative resistance. The term "negative resistance" refers to the occurrence of negative slope on part of the curve that represents current through the device as a function of voltage across it. Figure 8.3 shows such a curve. This kind of behavior can be produced by various mechanisms, one of which is emission of secondary electrons from an anode that is being bombarded by energetic electrons. Sometimes more electrons are emitted than are incident on the surface of the anode, and if there is an electrode available to collect these secondary electrons, the total current between cathode and anode may be diminished by an increase in the voltage difference between them. The increase of voltage causes the incident electrons to hit the anode with higher energy and thus to produce more secondaries. A tetrode vacuum tube sometimes operates in this way, with the so-called screen grid collecting the secondary electrons emitted from the anode. More generally, any device that amplifies can exhibit negative resistance if part of its output is applied to its input with appropriate sign to produce a reduction of current in the output as a result of an increase of voltage between input and output.

Before considering the effect of a voltage-current curve like that of Figure 8.3, let us first analyze the circuit in Figure 8.4. In

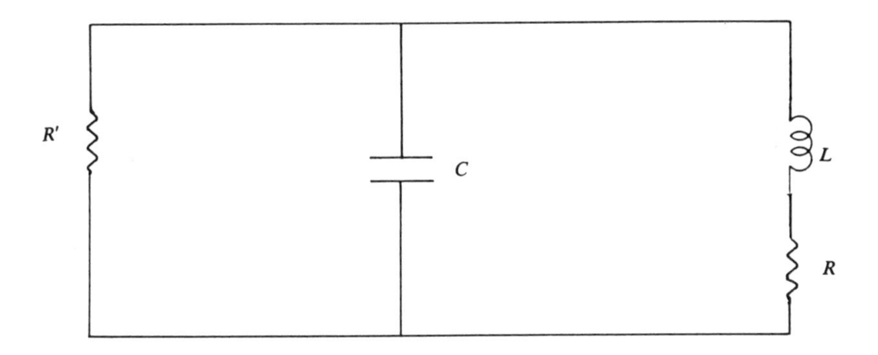

Figure 8.4 A circuit composed of linear elements. The resistance R' is assumed capable of having a negative value.

this circuit all the elements are assumed linear, so the circuit can be analyzed by use of linear differential equations for the loop currents:

$$\begin{aligned} R'\dot{I}_1 + (I_1 - I_2)/C &= 0; \\ (I_2 - I_1)/C + R\dot{I}_2 + L\ddot{I}_2 &= 0. \end{aligned} \tag{8.51}$$

Assuming a time-dependence e^{bt} for each current, one gets simultaneous linear equations for the complex amplitudes A_1, A_2:

$$\begin{aligned} (bR' + 1/C)A_1 - A_2/C &= 0 \\ -A_1/C + (1/C + bR + b^2L)\,A_2 &= 0 \end{aligned} \tag{8.52}$$

These equations are compatible if and only if the determinant of coefficients vanishes:

$$LR'b^3 + (RR' + L/C)b^2 + (R/C + R'/C)b = 0, \tag{8.53}$$

which has as roots

$$b = 0,$$

and

$$b = -\gamma \pm i\omega'. \tag{8.54}$$

The root $b = 0$ is spurious, a result of Equation 8.51 having already been differentiated with respect to t. In the other roots,

$$\gamma = \frac{R}{2L} + \frac{1}{R'C},$$

$$\omega' = \sqrt{\omega_0^2 - \gamma^2},$$

where (8.55)

$$\omega_0 = \frac{1}{\sqrt{LC}} \,.$$

Thus, if $\omega_0 > \gamma$, the undriven oscillations of the circuit appear to be like those studied in Chapter 2, with damping constant γ. But if we consider the possibility that $R' < 0$, we find that δ may be zero or negative. If

$$R' = -2L/RC, \tag{8.56}$$

there is no damping—the circuit oscillates with constant amplitude like one in steady-state response to a driving emf. If R' is more negative than this, the amplitude increases with time—a process that cannot continue forever.

It appears that a negative resistance can act to eliminate damping in a circuit that would otherwise dissipate energy. The negative resistance acts as a source of energy, as a positive resistance consumes energy. The physical object that has negative resistance thus contains, or is connected to, an energy source. In the case of negative resistance due to secondary electron emission, the energy source is the DC power supply that maintains the potential difference between anode and cathode.

A more realistic treatment of a device with negative resistance takes account of the kind of behavior represented in Figure 8.3, in which V is a multiple-valued function of I, and I is approximately expressible in the neighborhood of I_0 as

$$I = I_0 + (V - V_0)/R' + a(V - V_0)^3. \tag{8.57}$$

Because I is expressed explicitly as a function of V, it is convenient to analyze the circuit in Figure 8.5 by expressing the three branch currents in terms of E, the common potential difference across the three branches. We have

$$I + I_C + I_L = 0, \tag{8.58}$$

whence

$$\dot{I} + \dot{I}_C + \dot{I}_L = 0 = \frac{d}{dt}[I_0 + E/R' + aE^3 + CE + I_L]. \tag{8.59}$$

Now

$$E = RI_L + LdI_L/dt, \tag{8.60}$$

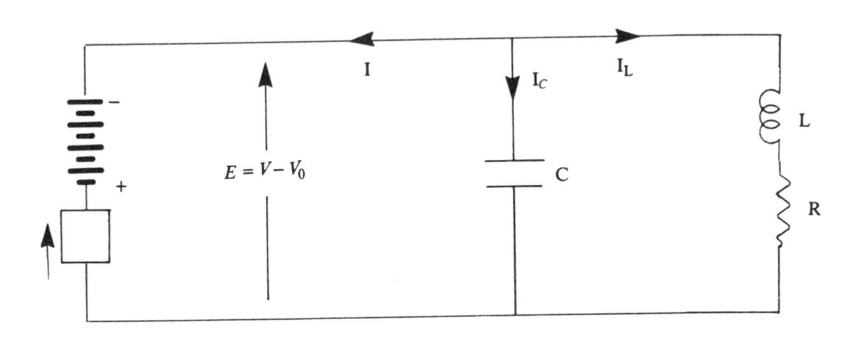

Figure 8.5. A more realistic depiction of the circuit shown in Figure 8.4. The box denotes the element with negative resistance.

so

$$dI_L/dt = E/L - (R/L)(-I-I_C). \tag{8.61}$$

Then Equation 8.59 becomes

$$\ddot{E} + (1/R'C + R/L + 3aE^2/C)\dot{E} + (1/LC + R/R'LC)E + (aR/LC)E^3 = 0. \tag{8.62}$$

In the case of most interest, $R' < 0$, as indicated in Figure 8.3. Looking at the case of very small R (interpretable as inductance with negligible resistance), we drop the terms containing R. Then, defining $\tau = r(LC)^{-1/2}$, we write

$$\frac{d^2E}{d\tau^2} + \left(\frac{L}{R'} + 3aLE^2\right)\frac{dE}{d\tau} + E = 0. \tag{8.63}$$

Finally we define

$$x = (-3aR')^{1/2}E = (3a\,|\,R'|\,)^{1/2}E; \qquad \varepsilon = -L/R' = L/|\,R'\,| > 0. \tag{8.64}$$

Then, denoting τ-derivatives by primes, we have

$$x'' - \varepsilon(1 - x^2)x' + x = 0. \tag{8.65}$$

This equation is known as the van der Pol equation, having been first studied in the 1920s by van der Pol in connection with self-sustaining oscillations. The qualitative behavior of x seems clear—when $x^2 < 1$, the damping is negative and the amplitude of oscillations increases with time, but when $x^2 > 1$, the damping is positive (more positive for larger x^2) so the amplitude decreases. One might expect a unique amplitude that is stable, neither increasing nor decreasing.

This argument is subject to the obvious objection that an oscillation of large amplitude covers values of x^2 both greater than and less than 1, so the argument about decreasing amplitude may be faulty. In order to get a new view of the problem, we consider orbits in the phase plane. We define $y = x'$, and write

$$
\left.\begin{aligned}
y' &= \varepsilon(1 - x^2)y - x, \\
x' &= y,
\end{aligned}\right\}
$$

so

$$
\frac{dy}{dx} = \varepsilon(1 - x^2) - \frac{x}{y}. \tag{8.66}
$$

For polar coordinates in the plane, we have

$$
\left.\begin{aligned}
r' &= \varepsilon\, r\sin^2\theta(1 - r^2\cos^2\theta), \\
\theta' &= \varepsilon\,\sin\theta\cos\theta(1 - r^2\cos^2\theta) - 1
\end{aligned}\right\}
$$

so

$$
\frac{dr}{r d\theta} = \frac{\varepsilon\sin^2\theta(1 - r^2\cos^2\theta)}{\varepsilon\,\sin\theta\cos\theta(1 - r^2\cos^2\theta) - 1}. \tag{8.67}
$$

Figure 8.6 shows the phase plane divided into regions according to the signs of r' and θ'.

The only equilibrium point (where $x' = y' = 0$) is the origin; it is obviously unstable, for $r' > 0$ there. So the system cannot remain at rest. As can be seen from Equation 8.65, damping is negative near the origin, that is, $r' > 0$ there, so the path in the phase plane, if followed clockwise, spirals outward in the inner region. Farther out, a path going clockwise around the origin passes through a region in which $r' > 0$ (near the y-axis) and another where $r' < 0$. It is not immediately obvious whether r increases or decreases in one trip around the origin, but by considering larger and larger initial value of r, one finds the length of the path in the region where $r' > 0$ approaching zero. The positive values of r' in this region are bounded, having ε as an upper limit. Therefore as $r \rightarrow \infty$, the increase of r near the y-axis approaches zero, whereas the decrease in other parts of the path has no such restriction. Thus very large paths spiral inward.

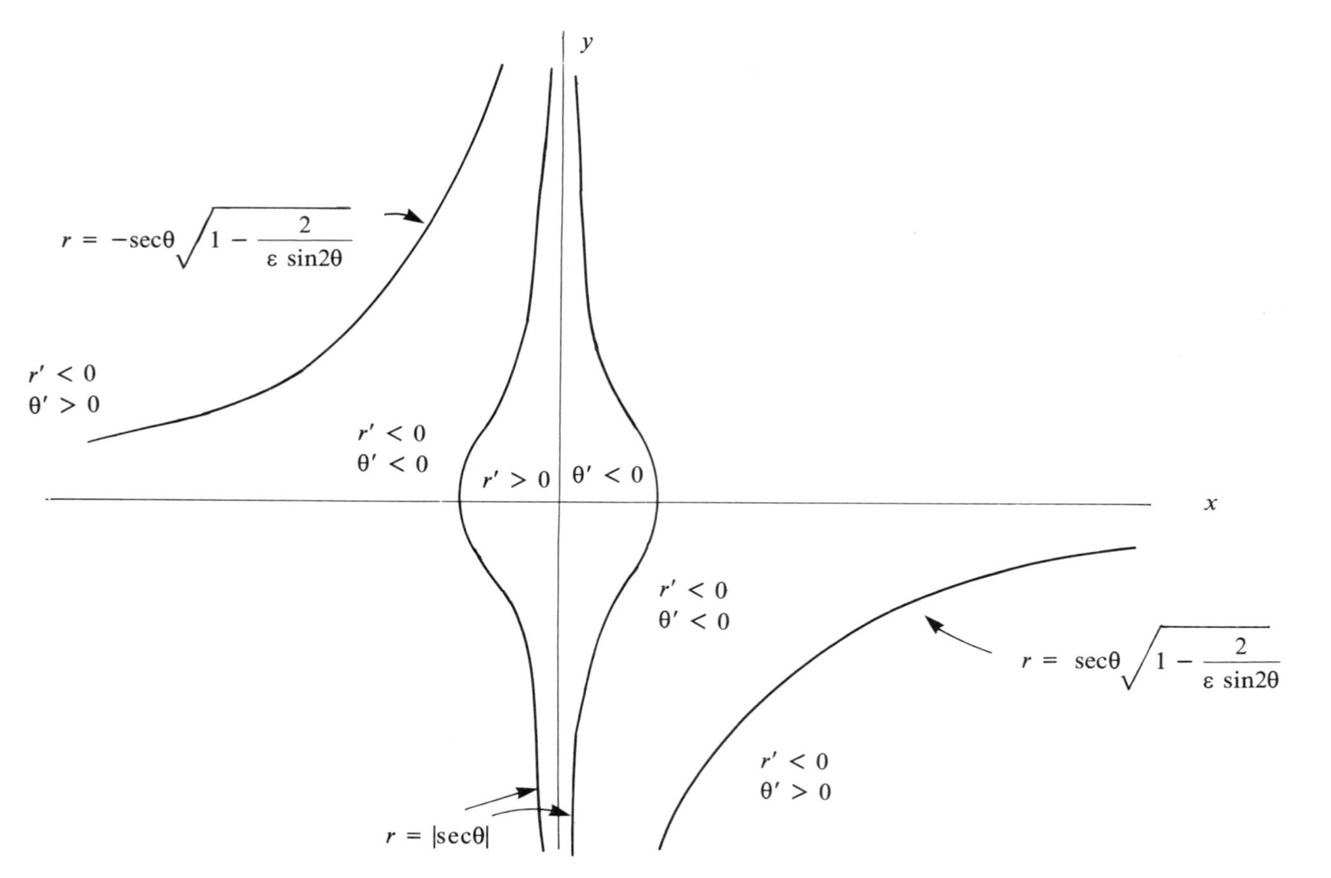

Figure 8.6. The phase plane for the van der Pol equation, with the signs of time derivatives in various parts of the plane indicated.

We have remarked that a point in phase space determines the system's motion. A consequence of this fact is that two paths in phase space can never cross (except at an equilibrium point); nor can a path cross itself. Therefore the fact that small paths spiral outward and large ones inward implies that, between these two extremes, there is at least one closed path toward which all others tend. The fact that both larger and smaller paths approach the closed one implies that the latter is stable; it is a limit cycle. This argument does not rule out the possibility of several limit cycles, but the shape of the $r' > 0$ region suggests that there will be a unique path on which the increase of r near the y-axis exactly balances its decrese in other regions. Whether there is one limit cycle or many, the system's behavior in a limit cycle is more complicated than in our previous examples, for there is now no cycle in which the damping force is everywhere zero. A limit cycle in the present case has positive damping on part of it and negative on the rest, so the two effects exactly balance out.

Merely knowing that a limit cycle exists does not tell us in detail what it is. In fact, it can be found only by a method of successive approximations. This method is easier to apply and more informative if used on the time-derivative equations instead of the path equation. Let us apply the method to Equations 8.66. The strategy, sometimes known as Picard's method, is to choose a first approximation to the function $x(\tau)$, called $x_0(\tau)$, and define $y_0(\tau) = x_0'(\tau)$. Then these functions are used to generate more accurate ones, which in turn produce others still better, and so on, according to the prescription

$$\begin{aligned} y_n' &= \varepsilon(1 - x_{n-1}^2)y_{n-1} - x_{n-1}, \\ x_n' &= y_n. \end{aligned} \tag{8.68}$$

Thus at each stage y determines x, and then the two together determine the y of the next stage. At each stage x and y are made to fit the initial conditions—each is expressed as an integral, to which a suitable constant is added for this purpose.

Usually the starting functions are taken to be solutions of the problem without damping. Thus possible starting functions are:

$$x_0 = a\cos\tau, \text{ so } x(0) = a,$$
$$y_0 = -a\sin\tau, \text{ so } y(0) = 0, \tag{8.69}$$

Then

$$y_1 = \int_0^\tau \tau 0\ [\varepsilon a \sin\tau(1 - a^2\cos 2\tau) - a\cos\tau]d\tau$$
$$= a\left[-\varepsilon\cos\tau - \sin\tau + \varepsilon\frac{a^2}{3}\cos^3\tau + \varepsilon\left(1 - \frac{a^2}{3}\right)\right]; \tag{8.70}$$
$$x_1 = a + \int_0^\tau y_1\, d\tau.$$

These equations already show a disease known as "secular terms." The constant term in y_1 represents a nonperiodic contribution to x' which, when integrated to give x_1, produces a term proportional to τ, a term that increases without limit as $\tau \rightarrow \infty$. These terms clearly do not belong in accurate expressions for x and y. In higher approximations such terms are even worse, for they become higher powers of τ, which behave worse at ∞. If the calculation could be continued through an infinite sequence of approximations, these terms might produce a power series in τ which would converge to a correct, well-behaved part of the solution, but in any finite sequence of terms they cause unacceptable behavior of x and y.

Because of this pathology, the foregoing calculation is not used on equations like the van der Pol equation. One that is used, which cannot be presented in available space, is to express x in the form

$$x(\tau) = a(\tau)\cos(\tau + \phi), \tag{8.71}$$

where a and ϕ are functions of τ that have to be determined by means of the equations of motion.

Although Picard's method is unsuited to the use outlined above, it gives somewhat better results when applied in polar coordinates. Now we write

$$r_n'/r_n = \varepsilon \sin^2\theta_{n-1}(1 - r_{n-1}{}^2 \cos^2\theta_{n-1}),$$

$$\theta_n' = -1 + \varepsilon \sin\theta_{n-1} \cos\theta_{n-1} (1 - r_{n-1}{}^2 \cos^2\theta_{n-1}), \qquad (8.72)$$

$$r_0(\tau) = \text{constant}; \ \theta_0(\tau) = -\tau.$$

From this starting point, we get as the next approximation

$$r_1 = r_0 \exp \tfrac{1}{4}\varepsilon[(2 - \tfrac{1}{2}r_0^2)\tau - \tfrac{1}{2}(2 - \tfrac{1}{4}r_0^2) \sin 2\tau];$$

$$(8.73)$$

$$\theta_1 = -\tau + \tfrac{1}{4}\varepsilon[3r_0^2/4 - 1 + (1 - \tfrac{1}{2}r_0^2) \cos 2\tau - \tfrac{1}{4}r_0^2 \cos^2 2\tau].$$

These functions also contain terms that look like secular terms, but in one case the term is in an exponent and in the other it is needed to make θ decrease monotonically as it should. It can be seen that the integrals that lead to r_2 and θ_2 are quite forbidding. Note however, that the expanding and contracting characteristic of paths is already present in these first-order solutions: r_1 is a periodic exponential times a fraction that increases if $r_0 < 2$ and decreases if $r_0 > 2$, and when $r_0 = 2$ it gives a description of the limit cycle (admittedly approximate). More accurate calculations verify this result, that the limit-cycle path intersects the x-axis at $x = 2$.

So far this chapter has dealt only with *autonomous* (i.e., isolated) systems, whose equations of motion contain no functions of t. The theory of nonlinear forced oscillations is, in comparison with the linear theory, complicated and fragmentary. Each different driving function leads to a new problem, because it is not possible to represent the effect of a sum of forces as the sum of their individual effects. We shall be able to do little more with this large set of problems than to extend our use of Picard's method in polar coordinates to the case in which there is a sinusoidal driving force. In the presence of such a driving force of frequency $\omega \equiv \alpha\omega_0$, the van der Pol equation becomes

$$x'' = \varepsilon(1 - x^2)x' - x + A \cos\alpha\tau + B \sin\alpha\tau. \qquad (8.74)$$

In terms of polar coordinates in the phase plane, this equation becomes

$$\frac{r'}{r} = \varepsilon \sin^2\theta(1 - r^2\cos^2\theta) + \frac{\sin\theta}{r}(A\cos\alpha\tau + B\sin\alpha\tau),$$

$$\theta' = -1 + \varepsilon\sin\theta\cos\theta(1 - r^2\cos^2\theta) + \frac{\cos\theta}{r} \tag{8.75}$$

$$\times (A\cos\alpha\tau + B\sin\alpha\tau).$$

Now we use the same starting point ("zero-order" approximation) as before: $r_0(\tau) =$ constant; $\theta_0(\tau) = -\tau$. Then we find that

$$\frac{r'_1}{r_1} = \varepsilon\sin^2\tau(1 - r_0^2\cos^2\tau) = \frac{\sin\tau}{r_0}(A\cos\tau + B\sin\delta\tau);$$

(8.76)

$$\theta_1' = -1 - \varepsilon\cos\tau\sin\tau(1 - r_0^2\cos^2\tau)\frac{\cos\tau}{r_0}(A\cos\alpha\tau + B\sin\tau);$$

Integrating these equations subject to the same initial conditions as before, $r(0) = r_0$, $\theta(0) = 0$, we get r_0

$$r_1 = r_0\exp\left\{\frac{\varepsilon}{4}[(2 - \tfrac{1}{2}r_0^2)\tau - \tfrac{1}{2}(2 - \tfrac{1}{4}r_0^2)\sin 2\tau]\right.$$

$$+ \frac{A}{2r_0}\left[\frac{\cos(1-\alpha)\tau - 1}{1-\alpha} + \frac{\cos(1+\alpha)\tau - 1}{1+\alpha}\right]$$

$$\left. - \frac{B}{2r_0}\left[\frac{\sin(1-\alpha)\tau}{1-\alpha} - \frac{\sin(1+\alpha)\tau}{1+\alpha}\right]\right\}; \tag{8.77}$$

$$\theta_1 = -\tau + \frac{\varepsilon}{4}\left[\frac{3r_0^2}{4} - 1 + (1 - \tfrac{1}{2}r_0^2)\cos 2\tau - \tfrac{1}{4}r_0^2\cos^2 2\tau\right]$$

$$+ \frac{A}{2r_0}\left[\frac{\sin(1-\alpha)\tau}{1-\alpha} + \frac{\sin(1+\alpha)\tau}{1+\alpha}\right]$$

$$- \frac{B}{2r_0}\left[\frac{\cos(1-\alpha)\tau - 1}{1-\alpha} + \frac{\cos(1+\alpha)\tau - 1}{1+\alpha}\right].$$

In the special case in which $\alpha = 1$, that is, the frequency of the applied force equals the natural frequency of the undamped oscillator, these quantities become

$$r_1 = r_0 \exp\left\{\frac{\varepsilon}{4}\left[(2 - \tfrac{1}{2} r_0^2)\tau - \tfrac{1}{2}(2 - \tfrac{1}{4} r_0^2)\sin 2\tau\right] + \frac{A}{4r_0}(\cos 2\tau - 1) + \frac{B}{2r_0}(\tau - \tfrac{1}{2}\sin 2\tau)\right\}$$

$$\theta_1 = -\tau + \frac{\varepsilon}{4}\left[\frac{3r_0^2}{4} - 1 + (1 - \tfrac{1}{2} r_0^2)\cos 2\tau - \tfrac{1}{4} r_0^2 \cos^2 2\tau\right] + \frac{A}{2r_0}(\tau + \tfrac{1}{2}\sin 2\tau) - \frac{B}{4r_0}(\cos 2\tau - 1). \tag{8.78}$$

In interpreting Equations 8.73, 8.77, and 8.78, it is necessary to remember that the coordinate x is given by

$$x = r\cos\theta. \tag{8.79}$$

Keep in mind, also, that the method of approximation used in deriving these equations depends on the correction to the initial choice of $r(\tau)$ and $\theta(\tau)$ being small, which usually means that ε and A and B are in some sense small.

If we write $\theta = -\tau + \phi$ in each case, then $\cos\theta$ is

$$\cos\theta = \cos\tau\cos\phi + \sin\tau\sin\phi. \tag{8.80}$$

Now the smallness of the difference between r_0, θ_0 and r_1, θ_1 implies that θ is small, so the functions of ϕ can be approximated:

$$\cos\theta_1 = \cos\tau + \phi\sin\tau. \tag{8.81}$$

A similar approximation in r_1 suggests replacing the exponential of trigonometric functions by the constant plus linear terms in the power series; the algebraic terms in the exponential are easy to interpret as they stand. Thus in each of the cases the expression for x contains products of $\sin\tau$ by the sines and cosines appearing in r and ϕ. Such products can be expressed as sines and cosines of sum and difference frequencies. In terms of t, the frequencies present in the undriven motion described in Equations 8.73 are multiples of ω_0, whereas in the nonresonant driven motion of Equations 8.77 there will be, in addition, ω and multiples of it. In the case of resonance, Equations 8.78, there is an additional monotonic motion of r, the direction of which depends on the sign of B, and an additional linear term in θ which will increase or decrease the frequency of the $\sin\tau$ term, depending on the sign of A. This change of frequency will generate shifts of the sum and difference frequencies, which will no longer be exact multiples of ω_0.

Because our approximation is too inaccurate when A and B are large, we cannot derive another effect that a more accurate calculation would reveal, the effect known as "entrainment." This means a rather abrupt shift, with increasing amplitude of the force, from periodic motion with the natural period of an oscillator to periodic motion with the period of the applied force. Such an effect may be predictable with use of a different $r_0(\tau)$ and $\theta_0(\tau)$ like those obtained by solution of the linear problem with $\varepsilon = 0$. The calculation using these initial functions is too laborious for inclusion here.

We conclude this sketchy discussion of nonlinear oscillations with a rather trivial observation. Some nonlinear equations can be reduced to manageable linear form by changes of dependent and independent variables—such changes can also remove explicit time-dependence from some equations, or make a damped system look as if it is undamped. A very simple example of this last effect is found in the change of the linear problem

$$x'' + 2sx' + x = f(\tau) \qquad (8.82)$$

by the substitution $x = ue^{-s\tau}$, which yields

$$u'' + (1 - s^2)u = f(\tau)e^{s\tau}. \qquad (8.83)$$

By hiding the effect of damping in the definition of the dependent variable, we have made the equation of motion look as if damping is absent.

More generally, if $y = g(x, t)$ and $T = f(t)$, the equation

$$\ddot{x} + \frac{g_{xx}}{g_x}\dot{x}^2 + \left[P(t) + \frac{2g_{xt}}{g_x}\right]\dot{x} + \frac{P(t)g_t + Q(t)g + g_{tt} - R(t)}{g_x} = 0 \tag{8.84}$$

can be converted to

$$\frac{d^2y}{dT^2} + 2\gamma\frac{dy}{dT} + \omega_0^2 y = F(\tau). \tag{8.85}$$

Here the subscripts denote partial derivatives with respect to the subscripted variables, and

$$Q(t) = \omega_0^2\left(\frac{dT}{dt}\right)^2;\ P(t) = \frac{2\gamma}{\omega_0}Q^{1/2} - \frac{\dot{Q}}{2Q};\ F(\tau) = \frac{\omega_0^2 R}{Q}. \tag{8.86}$$

The functions entering Equation 8.84 can be quite complicated and nonlinear, but one is unlikely to encounter an equation having precisely the relationships among its terms that are implied by the expressions in this equation.

Most of the work that has been done on nonlinear oscillations has been applied to one-dimensional systems. This chapter has summarized a small fraction of this work, for the purpose of showing why nonlinear problems are harder than linear ones, and of introducing some of the approaches that have been useful. Nonlinear many-dimensional systems without damping are well understood in principle (though not always easy to solve explicitly) by Lagrangian (see next chapter) and Hamiltonian methods. Those with damping are understood even less fully than damped one-dimensional nonlinear systems.

NOTES

All of these listed reference books in nonlinear oscillations have been used in the preparation of this chapter. The last two contain additional references to other books and papers on the subject.

Kryloff, N., and N. Bogoliuboff. *Introduction to Nonlinear Mechanics.* Princeton, NJ: Princeton University Press, 1943.

Butenin, N. N. *Elements of the Theory of Nonlinear Oscillations.* New York: Blaisdell, 1965

Meirovitch, L. *Elements of Vibration Analysis.* New York: McGraw Hill, 1975.

Mickens, R. E., *An Introduction to Nonlinear Oscillations.* New York: Cambridge University Press, 1981.

Lefschetz, S. *Differential Equations: Geometric Theory.* New York: Dover, 1977.

PROBLEMS

8.1. A particle moves in one dimension in the presence of a conservative force field. In each of the following cases, sketch the potential, work out the energy range that produces periodic motion, solve for $x(t)$, and determine the fundamental frequency of the motion.
(a) $U = -A\ \mathrm{sech}^2 kx$;
(b) $U = A \sec^2 kx$.

8.2. Discuss the motion of a system for which the function $g(r)$ introduced in Equation 8.33 is the reciprocal of a polynomial with real positive roots a, b, c . . . (in increasing order: $a < b < c \ldots$). In what circumstances does the system have limit cycles? Are there regions of the phase plane in which velocities are infinite? Does it matter whether all the roots are single, or some are double or multiple?

8.3. Apply the first correction calculated by Picard's method to the undriven linear damped oscillator, taking the undamped motion as the starting solution and then correcting this solution in one step. What errors are present in the improved $x(\tau)$?

9

Coupled Oscillators without Damping—Lagrange's Equations

Having discussed electrical networks in rather general terms, we might appear to be needlessly repeating now when we begin a new discussion of coupled mechanical oscillators without damping. Indeed, the systems that now interest us are a subset of those treated previously, but we shall discuss them in a rather different way. Instead of asking for the response of a system to a given force $F(t)$, we shall attend mainly to the undriven motion of a set of coupled undamped oscillators, and our principal interest will be the ways in which such a system can move in simple harmonic motion. These ways of moving, known as *normal modes* of motion, have their own characteristic *normal frequencies;* they are important in various application of quantum theory—in particular, the theory of molecular vibrations and the theory of vibrations of crystal lattices, and, by extension, the quantum theory of fields. Although we shall not be applying quantum theory, we can in the context of classical mechanics carry out the part of such calculations that is common to classical and quantum theory. The mathematical methods and the physical ideas encountered in this work are much the same in classical as in quantum theory.

In order to have useful equations of motion, we shall first develop Lagrange's form of these equations, as they apply to a nondissipative system. This form has the great advantage of being applicable in all coordinate systems, provided that the kinetic and potential energies are written in terms of the appropriate coordinates and their time derivatives. We begin by writing Newton's equations for a system of N particles:

$$m_k \ddot{x}_k = -\partial V/\partial x_k + \phi_k, \tag{9.1}$$

where x_k means x, y, or z for one of the particles—$k = 1, 2, 3, \ldots, 3N$, for a system of N particles—and m_k is the mass of the particle which has x_k as one of its coordinates, $m_1 = m_2 = m_3$, etc.—and V is the potential energy of the system, a function of the x_k but independent of the $\dot{x}_k$—and ϕ_k is an applied force, assumed known. Now we contemplate a change of coordinates from the x_k to a set $q_1, q_2, \ldots, q_{3N}$. The transformation can be defined by a set of equations whereby the old coordinates x are expressed as functions of the new ones, q. We multiply Equation 9.1 by the partial derivatives of x_k with respect to q_n, where q_n is any one of the new coordinates, and sum over k:

$$\sum_k m_k \ddot{x}_k \frac{\partial x_k}{\partial q_n} = -\sum_k \frac{\partial V}{\partial x_k}\frac{\partial x_k}{\partial q_n} + \sum_k \phi_k \frac{\partial x_k}{\partial q_n} = -\frac{\partial V}{\partial q_n} + F_n; \tag{9.2}$$

$$\begin{aligned}\sum_k m_k \ddot{x}_k \frac{\partial x_k}{\partial q_n} &= \frac{d}{dt}\sum_k m_k \dot{x}_k \frac{\partial x_k}{\partial q_n} - \sum_k m_k \dot{x}_k \frac{d}{dt}\frac{\partial x_k}{\partial q_n} \\ &= \frac{d}{dt}\sum_k m_k \dot{x}_k \frac{\partial \dot{x}_k}{\partial \dot{q}_n} - \sum_k m_k \dot{x}_k \frac{\partial \dot{x}_k}{\partial q_n},\end{aligned} \tag{9.3}$$

where the second sum has been transformed by reversal of the order of differentiation, and the first by the relations

$$\dot{x}_k = \sum_n \frac{\partial x_k}{\partial q_n}\dot{q}_n + \frac{\partial x_k}{\partial t},$$

whence

$$\frac{\partial \dot{x}_k}{\partial \dot{q}_n} = \frac{\partial x_k}{\partial q_n}. \tag{9.4}$$

Thus, from Equation 9.2 and 9.3, we find that, if T is kinetic energy

$$\left(\frac{d}{dt}\frac{\partial}{\partial \dot{q}_n} - \frac{\partial}{\partial q_n}\right)\sum_k \tfrac{1}{2}m_k\dot{x}_k^2 \equiv \frac{d}{dt}\frac{\partial T}{\partial \dot{q}_n} - \frac{\partial T}{\partial q_n} = -\frac{\partial V}{\partial q_n} + F_n, \tag{9.5}$$

or

$$\frac{d}{dt}\frac{\partial L}{\partial \dot{q}_n} - \frac{\partial L}{\partial q_n} = F_n, \tag{9.6}$$

where the *Lagrangian function* L is given by

$$L = T - V. \tag{9.7}$$

Equation 9.6 is known as *Lagrange's equation*—more properly, Lagrange's equation*s*, since there is one equation for each choice of the index n. We assumed that we were dealing with N particles in three dimensions, hence with $3N$ Cartesian coordinates x_k; thus there are equally many of the q_n. However, there are likely to be constraints acting in the system—in particular, so-called *holonomic constraints,* which are algebraic relationships among the coordinates such as the requirement that a particle moves on the surface of a given sphere, or on a tabletop, or at a given distance from another particle. When there are constraints, they are associated with forces, known as *forces of constraint,* whose action causes the constraints to be in effect.

The tension in the string that supports a pendulum bob is an example of a force of constraint. Such forces are not known in advance, for they have whatever values and directions they need to have in order to enforce the constraints; these forces are known initially solely by their effects (e.g., the bob remains always at the same distance from the point of support). Such forces are not known as forces until the problem of motion has been solved (e.g., one knows the tension in the string only after learning how the pendulum moves). Thus, in order to make progress in solving a problem that involves holonomic constraints, one imposes them as restrictions on the values of the coordinates without asking about the forces of constraint. So whatever part of V in Equation 9.1 gave rise to these unknown forces, this part of V should be excluded from the Lagrangian function $T - V$, and its effect should be embodied in the solution of the equations of motion. The easiest way to take account of such constraints is to choose coordinates appropriately, letting each constraint equation simply state that one of the coordinates q_n has a given constant value. For example, in the case of the pendulum, choose one of the coordinates to be the distance r from the bob to the point of support, and the constraint then amounts to r having the constant value (the length of the string). Then, by virtue of the constraints, some of the equations of motion are already solved and fewer than $3N$ remain to be solved.

A possible obstacle to this procedure would be forces of constraint that do more than enforce certain algebraic relationships, e.g., friction between a mass and a surface on which it is constrained to move. In such cases the equations of motion ought to contain extra terms representing these forces in their effect on the unconstrained coordinates—and such terms would be impossible to write before the forces of constraint were known. We shall assume that no such forces are present, that all forces of constraint are perpendicular to the curves and surfaces on which objects are constrained to move. This assumption is conventionally embodied in d'Alembert's principle—that forces of constraint do no work.[1]

Taking our equations of motion to be in the form of Equation 9.6 with $F_n = 0$, we note that this form takes account of all forces that can be derived from the potential-energy function V. The definition of V can be generalized to include some velocity-dependent forces, such as magnetic forces on charged particles, but dissipative forces cannot be included in the form that we are using. Such forces, whether they are caused by constraining objects or by something else, require the inclusion in Lagrange's equations of extra terms. In assuming that our systems have neither frictional nor driving forces, we are assuming that such extra terms are absent; if there are explicitly given driving forces, they can be included as a set of known functions $F_n(t)$.

The partial derivative $\partial L/\partial \dot{q}_n$ is given a special name and a special symbol:

$$\frac{\partial T}{\partial q_n} = \frac{\partial L}{\partial \dot{q}_n} \equiv p_n, \tag{9.8}$$

the *momentum* conjugate to q_n. In these terms, Equation 9.6 becomes

$$\dot{p}_n = \frac{\partial L}{\partial q_n}. \tag{9.6'}$$

Before going on to use Lagrange's equations, we digress to establish a result that we shall use in a later chapter. Let us take Equation 9.6, with $F_n = 0$, to contain only unconstrained coordinates q_n, the others having been assigned constant values. Let us now choose a set of differentiable functions of t, $g_n(t)$, which are all required to vanish at two particular times t_1 and t_2:

$$g_n(t_1) = g_n(t_2) = 0. \tag{9.9}$$

Now let us multiply each of the Equations 9.6 by one of these functions, add all the equations together, and integrate with respect to t from t_1 to t_2:

$$\int_{t_1}^{t_2} \sum_n \left(\frac{d}{dt} \frac{\partial L}{\partial \dot{q}_n} - \frac{\partial L}{\partial q_n} \right) g_n(t)\, dt = 0. \tag{9.10}$$

Now let us integrate by parts the first term with each value of n, and use Equation 9.9 to get rid of the uv contribution,

$$-\int_{t_1}^{t_2} \sum_n \left(\frac{\partial L}{\partial \dot{q}_n} \dot{g}_n(t) + \frac{\partial L}{\partial q_n} g_n(t) \right) dt = 0. \tag{9.11}$$

This integral has a direct and simple interpretation. Let us take the functions $q_n(t)$ to be the solution of the equations of motion,

9.6. Then let us define a set of functions $Q_n(t)$ to represent a varied motion that still conforms to the constraints, if any, but does *not* satisfy the equations of motion; let us require the varied motion to coincide with the actual motion at times t_1 and t_2:

$$Q_n(t_1) = q_n(t_1);\ Q_n(t_2) = q_n(t_2). \tag{9.12}$$

This state of affairs can be represented by

$$Q_n(t) = q_n(t) + hg_n(t), \tag{9.13}$$

where h is a real constant—$h = 0$ yields the actual motion—and the $g_n(t)$ are the functions introduced before Equation 9.9. Then, L being a function of the Q_n, the $\dot{Q}_n$ (and perhaps of t explicitly) a priori (when the motion is not required to satisfy Equation 9.6), the integrand of Equation 9.11 can be identified as the derivative of L with respect to h evaluated at $h = 0$:

$$\begin{aligned} \frac{dL}{dh}\Big|_{h=0} &= \sum_n \left[\frac{\partial L}{\partial \dot{Q}_n}\frac{\partial \dot{Q}_n}{\partial h} + \frac{\partial L}{\partial Q_n}\frac{\partial Q_n}{\partial h}\right] \\ &= \sum_n \left[\frac{\partial L}{\partial \dot{q}_n}\dot{g}_n + \frac{\partial L}{\partial q_n}g_n\right] \end{aligned} \tag{9.14}$$

Thus we have shown that Lagrange's equations imply the vanishing of the time integral of this derivative, regardless of the particular functions g_n that are used, provided they are differentiable and satisfy Equation 9.9:

$$\int_{t_1}^{t_2} dL/dh\,|_{h=0}\,dt = 0 = d/dh \int_{t_1}^{t_2} Ldt\,|_{h=0}, \tag{9.15}$$

where the last form results from the fact that the limits t_1, t_2 do not depend on h. This form is often written as

$$\delta \int_{t_1}^{t_2} L\,dt = 0. \tag{9.16}$$

The statement 9.16 means the last statement of 9.15; it is known as *Hamilton's principle,* or sometimes as the principle of least action. It states that, of all the motions that are conceivable between fixed configurations at fixed times, that motion will actually occur that gives $\int L dt$ a stationary value—usually a minimum.

We have shown that Hamilton's principle follows from Lagrange's equations. It is no harder to show that Lagrange's equations follow from Hamilton's principle. Hamilton's principle is an alternative formulation of the laws of motion, equivalent to the other formulations. On account of its simplicity and its independence of particular coordinate systems, it is sometimes the most convenient form of the laws of motion. For the present, however, we shall use the form in Equation 9.6.

We wish to apply Equation 9.6 to a system that has a configuration of stable equilibrium, in order to study small vibrations around the equilibrium configuration. Thus, as in Chapter 1, we contemplate Taylor series taken about the equilibrium configuration and drop all terms of higher degree than second in the displacements from equilibrium and the time derivatives of these displacements. Aside from the greater number of coordinates, what is different now from the treatment in Chapter 1 is the explicit involvement of kinetic energy T in the equations of motion. T is commonly given by an expression of the form

$$T = \tfrac{1}{2} \sum_{mn} G_{mn}(q)\, \dot{q}_m \dot{q}_n, \tag{9.17}$$

whence

$$p_n \equiv \frac{\partial T}{\partial \dot{q}_n} = \sum_m G_{mn} \dot{q}_m, \tag{9.18}$$

and

$$\dot{p}_n = \sum_m G_{mn} \ddot{q}_m + \sum_{mk} \frac{\partial G_{mn}}{\partial q_k} \dot{q}_m \dot{q}_k.$$

This equals $\partial L/\partial q_n$, which is

$$\partial L/\partial q_n =$$

$$\partial T/\partial q_n - \partial V/\partial q_n = \tfrac{1}{2}\sum_{mk} (\partial G_{mk}/\partial q_n)\dot{q}_m \dot{q}_k - \frac{\partial V}{\partial q_n}. \qquad (9.19)$$

Bringing all these terms together and (as is sometimes done) writing the double sum in p_n in a form that makes its symmetry more obvious, we find that

$$\sum_m G_{mn} \ddot{q}_m + \tfrac{1}{2}\sum_{mk}\left(\frac{\partial G_{mn}}{\partial q_k} + \frac{\partial G_{kn}}{\partial q_m} - \frac{\partial G_{mk}}{\partial q_n}\right)\dot{q}_m\dot{q}_k = -\frac{\partial V}{\partial q_n}, \qquad (9.20)$$

whence we can see that equilibrium (all time derivatives zero) requires that all the $\partial V/\partial q_n$ also be zero—still the criterion of equilibrium. Assuming that there is a set of q values (a configuration) for which all these derivatives vanish, we let s_k be the excursion of q_k from its equilibrium value, and expand V in Taylor series about equilibrium; here we abbreviate the set $s_1, s_2, \ldots, s_N$ as simply s:

$$V(s) = V(0) + \sum_k s_k V_k(0) + \tfrac{1}{2}\sum_{kj} s_k s_j V_{kj}(0) + \ldots, \qquad (9.21)$$

where the subscripts on V denote partial derivatives with respect to the indicated coordinates, and the (0) means that these derivatives are evaluated at equilibrium. Then, as previously, the linear terms all vanish because the first derivatives are zero at equilibrium, by definition. If the excursions from equilibrium, s, are small, we can neglect cubic and higher terms. We shall drop $V(0)$, also, because such a constant term in potential energy can be chosen for convenience. So, now omitting the (0) where it is no longer necessary, we have

$$V = \tfrac{1}{2}\sum_{k,j} V_{kj}\, s_k s_j. \qquad (9.22)$$

Here the V_{kj} are a set of numerical constants which, by their origin as second derivatives, satisfy

$$V_{kj} = V_{jk}. \qquad (9.23)$$

Now by Equation 9.17 the kinetic energy T can be written as

$$T = \tfrac{1}{2} \sum_{m,n} G_{mn}(s)\, \dot{s}_m \dot{s}_n \tag{9.24}$$

and the G_{mn} can be expanded in Taylor series about the equilibrium configuration (where there is no reason to suppose that any derivatives of the Gs vanish); thus the G_{mn} will be expressed as power series in the s. However, in keeping with our policy of discarding terms of higher than second degree in the s and the $\dot{s}$, we drop all terms containing derivatives of the G_{mn}, and indicate by T_{mn} the value of G_{mn} at equilibrium. Then, relabeling some dummy variables, we have

$$L = T - V = \tfrac{1}{2}\sum_{m,n} T_{mn}\dot{s}_m \dot{s}_n - \tfrac{1}{2}\sum_{m,n} V_{mn} s_m s_n, \tag{9.25}$$

where we can take $T_{mn} = T_{nm}$ as a consequence of $G_{mn} = G_{nm}$, which could be assumed in Equation 9.17 with no loss of generality. Now, with L assumed quadratic in the s and the $\dot{s}$ and all F_n assumed absent, we derive from 9.6 the equations of motion

$$\sum_n T_{mn}\ddot{s}_n + \sum_n V_{mn} s_n = 0. \tag{9.26}$$

These equations bear some resemblance to the equation of motion of an undamped oscillator, but they are coupled. Each equation (for each choice of m) contains many of the $\ddot{s}$ and many of the s. Until these equations can be uncoupled, we cannot learn much about our system's small vibrations about equilibrium. Therefore, in the next chapter, we shall discuss the use of matrix algebra in uncoupling sets of simultaneous equations such as Equation 9.26.

NOTES

1. For more discussion of d'Alembert's principle and Lagrange's equations in general, see Goldstein, H. *Classical Mechanics.* 2nd ed. Reading, MA: Addison Wesley, 1980.
2. Lagrange's equations are also discussed more or less fully and used

by many other authors. See, for example, Symon, K.R. *Mechanics*. 3rd. ed. Reading, MA: Addison Wesley, 1971.

PROBLEMS

9.1 Write a Lagrangian and derive differential equations of motion for one particle in a conservative force field in three dimensions, using the following coordinate systems:
(a) Cartesian,
(b) Cylindrical,
(c) Spherical.

9.2 Choose convenient coordinates that permit separation of center-of-mass from relative motion of two particles subject only to a conservative central force that they exert on each other. Then write a Lagrangian and derive differential equations of motion.

9.3 For each of the potentials in Problem 8.1 work out the frequency of small oscillations. Is this greater or less than the fundamental frequency calculated in Problem 8.1?

10

Matrices—Rotations—Eigenvalues and Eigenvectors—Normal Coordinates

In the preceding chapter we encountered the problem of uncoupling the simultaneous equations of motion arising from the Lagrangean

$$L = T - V = \frac{1}{2}\sum_{m,n} T_{mn}\dot{s}_m\dot{s}_n - \frac{1}{2}\sum_{m,n} V_{mn}s_m s_n. \tag{10.1}$$

We shall attempt to do this by means of a linear, homogeneous transformation of coordinates:

$$s_n = \sum_j W_{nj}\, z_j \tag{10.2}$$

which converts L to the form

$$L = \frac{1}{2}\sum_k M_k\dot{z}_k^2 - \frac{1}{2}\sum_k K_k z_k^2, \tag{10.3}$$

whence the equations of undriven motion (all $F_n = 0$) are

$$M_k\ddot{z}_k + K_k z_k = 0, \tag{10.4}$$

so that each of the new coordinates z_k undergoes simple harmonic motion with frequency $\omega_k = (K_k/M_k)^{1/2}$. These are the normal frequencies, and the z_k are the *normal coordinates* of the system.

In seeking to calculate the normal coordinates of a system, we shall use some of the methods of matrix algebra (which subject we shall therefore expound as needed). A matrix is a rectangular array of numbers (real numbers in the cases that will concern us) which is treated as a single mathematical object. The numbers are arranged in rows and columns; a matrix is called $m \times n$ if it has m rows and n columns. The numbers in a matrix are commonly designated by the symbol that denotes the whole matrix, with two subscripts; the first subscript indicates which row the particular number belongs in, and the second subscript denotes the column. Thus A_{13} is the number at the intersection of the first row and the third column in the matrix A. The numbers that make up a matrix are known as the *elements* of the matrix.

Matrices that have the same size and shape (that is, the same number of rows and the same number of columns) are said to be *similar*. Two similar matrices are *equal* if all their corresponding elements are respectively equal. Nonsimilar matrices cannot be equal. *Addition* and *subtraction* are defined only for similar matrices; they simply involve adding (or subtracting) all corresponding elements of the two matrices. Thus,

$$(A + B)_{ij} = A_{ij} + B_{ij}. \tag{10.5}$$

Clearly addition, so defined, is commutative and associative:

$$A + B = B + A,$$

and (10.6)

$$(A + B) + C = A + (B + C) = A + B + C.$$

To multiply a matrix by a number, one multiplies all its elements by the number:

$$(NA)_{ij} = NA_{ij}. \tag{10.7}$$

Then subtraction of matrices can be defined in terms of addition:

$$A - B = A + (-1)B;$$
$$(A - B)_{ij} = A_{ij} - B_{ij}. \tag{10.8}$$

The *zero* matrix with m rows and n columns is the $m \times n$ matrix containing all zeros. This matrix, when added to any matrix similar to it, leaves the latter unchanged.

Multiplication of matrices is defined as follows, in terms of elements:

$$(AB)_{ij} = \sum_{n} A_{in} B_{nj}. \tag{10.9}$$

This process is defined only when the number of columns in the first factor equals the numbers of rows in the second; otherwise, the index n does not have a clearly defined range. The product of an $m \times n$ matrix with an $n \times p$ matrix is then an $m \times p$ matrix. It is not difficult to show that multiplication is in general noncommutative, but it is associative. Also it is distributive with respect to addition:

$$AB \neq BA \text{ except in special cases;}$$
$$(AB)C = A(BC); \tag{10.10}$$
$$A(B + C) = AB + AC; (B + C)A = BA + CA.$$

We shall be concerned with $n \times n$ matrices, known as *square* matrices, with $n \times 1$ matrices, known as *column* matrices (n rows and 1 column), and $1 \times n$ *row* matrices. We can schematically indicate which products of such matrcies exist and what they are like:

a $(n \times n)(n \times n) = (n \times n)$;
b $(n \times n)(n \times 1) = (n \times 1)$;
c $(n \times n)(1 \times \mathrm{n})$ undefined;
d $(n \times 1)(n \times n)$ undefined;

e $(n\times 1)(n\times 1)$ undefined; (10.11)
f $(n\times 1)(1\times n) = (n\times n)$, a rare case;
g $(1\times n)(n\times n) = (1\times n)$;
h $(1\times n)(n\times 1) = (1\times 1)$, a single number;
i $(1\times n)(1\times n)$ undefined.

Cases a, b, g, and h are the ones that we shall encounter.

In the set of $n \times n$ square matrices there is one known as the *unit* matrix I; its elements are

$$\begin{aligned} I_{mn} = \delta_{mn} &= 0 \text{ if } m \neq n, \\ &= 1 \text{ if } m = n. \end{aligned} \tag{10.12}$$

I is an example of a *diagonal* matrix, i.e., a square matrix whose off-diagonal elements are all zero. Using Equation 10.9, one can easily verify that multiplying any matrix by I leaves the matrix unchanged.

The *transpose* of a matrix A, which we shall designate $\tilde{A}$ (some write, instead, A^T), is obtained from A by interchange of rows and columns:

$$(\tilde{A})_{ij} = A_{ji}. \tag{10.13}$$

This process changes an $(m\times n)$ matrix into an $(n\times m)$ matrix. With the aid of Equation 10.9, it can be easily verified that the transpose of a product of matrices equals the product of transposes of the factors, multiplied in opposite order. For example,

$$(\widetilde{AB}) = \tilde{B}\,\tilde{A}; \tag{10.14}$$

the same rule applies also to products of more than two factors.

The *reciprocal* of a matrix, if it has one, is designated by an exponent -1 (e.g., A^{-1}); it is so defined that

$$AA^{-1} = A^{-1}A = I, \tag{10.15}$$

the unit matrix of appropriate size. We define the reciprocal only for square matrices, and not all of them have reciprocals; those that do not are called *singular*. Whether A is singular depends on whether Equation 10.15, written out in terms of components, can be solved for the unknown elements of A^{-1}. A little thought convinces us that a set of solutions exists if and only if the *determinant* of A is nonzero. Thus a singular matrix is one that has a zero determinant.

Another theorem that we shall cite without proof[1] is that the determinant of a product of similar square matrices is the product of the determinants of the separate matrices. Thus, for example, a product containing one singular factor is itself singular.

From the definition of the reciprocal of a matrix it follows directly that the reciprocal of a matrix product is the product of the individual reciprocals in reversed order, for example,

$$(AB)^{-1} = B^{-1}A^{-1}. \tag{10.16}$$

A matrix that is equal to its own transpose is known as *symmetric:*

$$S = \tilde{S}. \tag{10.17}$$

It is *antisymmetric* if it is the negative of its own transpose:

$$A = -\tilde{A}. \tag{10.18}$$

A matrix whose transpose equals its reciprocal,

$$\begin{aligned} \tilde{R} &= R^{-1}, \\ \tilde{R}R &= R\tilde{R} = I, \end{aligned} \tag{10.19}$$

is known as *orthogonal.* These equations combined with the theorem about determinants of products imply that the determinant of an orthogonal matrix (equal to the determinant of its transpose) is plus or minus 1—since its square equals the determinant of the unit matrix I, namely, plus 1.

Matrices are commonly interpreted in terms of vector spaces. We are working in an N-dimensional vector space over the real numbers—N-dimensional because N is defined as the number of coordinates in our mechanical system—"over the real numbers" because our vectors all have only real components. In such a space, vectors are represented as column or row matrices, and square matrices are usually regarded as linear operators that change vectors into other vectors. A column matrix s

$$s = \begin{pmatrix} s_1 \\ s_2 \\ \cdot \\ \cdot \\ s_N \end{pmatrix}$$

and the row matrix $\tilde{s}$

$$\tilde{s} = (s_1 \; s_2 \; \ldots \; s_N),$$

containing the same numbers as s, are usually taken to represent the same vector, a vector with components s_1, s_2, etc.

In our problem, these components might be the x, y, and z components of the displacements of the particles in the system from equilibrium; that is what the s_i mean in Equation 10.1. Or they might be the corresponding velocity components $\dot{s}_i$—also as in Equation 10.1. In either case (let us take the coordinates s_i as an example) the vector whose components are the s_i can be thought of as a vector in an N-dimensional space known as *configuration space.* Such a vector in this space determines and represents the entire configuration of the system, that is, the position of every particle in the system.

As in more familiar examples, a given vector can be resolved into components in various ways, depending on the choice of coordinate axes. Thus the configuration vector can be expressed in terms of components different from s_i. Such a change of axes in configuration space (which amounts to a coordinate transformation in the mechanical problem) can most easily be discussed in terms of base vectors. We assume that there are two sets of N orthogonal unit vectors (like $\mathbf{i}, \mathbf{j}, \mathbf{k}$ of familiar three-space). Let us call the sets $\mathbf{e}_i$ and $\mathbf{f}_i$ ($i = 1, 2, \ldots, N$). Let us assume that the s_i are the components of the configuration vector referred to the $\mathbf{e}_i$ vectors, and let us recognize that the same vector can be referred to the $\mathbf{f}_i$, with other components z_i:

$$\sum_i \mathbf{e}_i s_i = \sum_i \mathbf{f}_i z_i. \tag{10.20}$$

Now (compare Equations 4.14–4.16) we can isolate one s_i and thus express it in terms of the z_i by multiplying Equation 10.20 by one of the $\mathbf{e}$ vectors, say $\mathbf{e}_n$:

$$\mathbf{e}_n \cdot \sum_i \mathbf{e}_i s_i = s_n = \sum_i (\mathbf{e}_n \cdot \mathbf{f}_i)\, z_i = \sum_i R_{ni} z_i, \tag{10.21}$$

or, in terms of the column matrices s and z, and the square matrix R,

$$s = Rz. \tag{10.22}$$

In Equation 10.21, we used the assumed orthogonality and unit length of the $\mathbf{e}_i$ vectors. The square matrix R obviously has the elements

$$R_{ni} = \mathbf{e}_n \cdot \mathbf{f}_i. \tag{10.23}$$

Note that, with the help of Equation 10.14, we can rewrite Equation 10.22 as

$$\tilde{s} = \tilde{z}\tilde{R}. \tag{10.24}$$

Note, also, that we can multiply Equation 10.20 by one of the $\mathbf{f}$ vectors, thus deriving the inverse transformation equations:

$$z_m = \sum_i (\mathbf{f}_m \cdot \mathbf{e}_i)\, s_i = \sum_i R_{im} s_i = \sum_i \tilde{R}_{mi} s_i, \tag{10.25}$$

where we have used the definition 10.23 of the elements of the matrix R. Thus we have, in terms of whole matrices,

$$z = \tilde{R}s,$$

and (10.26)

$$\tilde{z} = \tilde{s}\,R.$$

Now, using Equation 10.22 and associativity, we have

$$z = \tilde{R}s = \tilde{R}Rz. \tag{10.27}$$

Since this result does not depend on the matrices s and z containing any particular set of numbers but must be true for all such sets of numbers, the matrix R by itself must satisfy

$$\tilde{R}R = I, \tag{10.28}$$

or R is orthogonal. So it appears that, in a real vector space, the transformation from an orthonormal basis to another orthonormal basis is effected by an orthogonal transformation matrix.

The converse, that each orthogonal matrix transforms from a given defined orthonormal basis to some other orthonormal basis, is established with the help of the invariance of scalar products. Let two vectors, referred to the **e** basis, be multiplied:

$$(\sum_i s_i \mathbf{e}_i) \cdot (\sum_i s_j{}' \mathbf{e}_j) = \sum_i s_i s_j{}'(\mathbf{e}_i \cdot \mathbf{e}_j) = \sum_i s_i s_i{}' = \tilde{s}s', \tag{10.29}$$

where we have used the orthonormality of the **e** vectors and, furthermore, can see that, if every dot product of two vectors equals the matrix product given in the last step, the basis *must* be orthonormal. Now, if s and s' transform under Equations 10.22 and 10.24, we get

$$\tilde{s}s' = \tilde{z}\tilde{R}Rz' = \tilde{z}z', \tag{10.30}$$

where the orthogonality of R has been used in the last step. But $\tilde{s}s'$ has been shown to be a dot product, a unique number that must be the same when the vectors are expressed in a new coordinate system. Therefore $\tilde{z}z'$ equals this same dot product, which establishes that the system is orthonormal—otherwise the dot product would not be $\tilde{z}z'$.

Let us now note that the Lagrangian in Equation 10.1 can be written, in matrix notation, as

$$L = \tfrac{1}{2}\tilde{\dot{s}}T\dot{s} - \tfrac{1}{2}\tilde{s}Vs \tag{10.1'}$$

where the matrices T and V are the real symmetric matrices with elements equal to the T_{mn} and V_{mn} of 10.1. Note, also, that 10.3 can be written

$$L = \tfrac{1}{2}\tilde{\dot{z}} M\dot{z} - \tfrac{1}{2}\tilde{z}Kz, \tag{10.3'}$$

where M and K are diagonal matrices (matrices containing no other nonzero elements except $M_{kk} = M_k$ and the $K_{kk} = K_k$). We have posed to ourselves the problem of finding a transformation matrix W, to be used as in Equation 10.2, that will change L from the form 10.1′ to the form 10.3′.

We shall first deal with a special case of the problem, in which the matrix T already has the desired form M, and only V has to be changed (without the diagonal form of T being affected). This special case is quite common. If the s are simply the Cartesian coordinates of particles' displacements from equilibrium, then

$$T = \sum_i \tfrac{1}{2} m_i\dot{s}_i^2 = \tfrac{1}{2}\tilde{\dot{s}} M\dot{s}, \tag{10.31}$$

where the diagonal matrix M has as its nonzero elements M_k simply the masses of the individual particles. Our task at present is to guarantee that this form will be preserved when the transformation is applied. One way to make sure of this is to change the scale on all the axes in configuration space, so that the matrix M gets replaced by the identity I:

$$\begin{aligned} x_i &= m_i^{1/2} s_i; \\ \dot{x}_i &= m_i^{1/2} \dot{s}_i. \end{aligned} \tag{10.32}$$

Defining

$$U_{ij} = V_{ij}(m_i m_j)^{-1/2}, \tag{10.33}$$

we note that, like V, U is real and symmetric. Then

$$L = \tfrac{1}{2}\tilde{\dot{x}}\dot{x} - \tfrac{1}{2}\tilde{x}Ux. \tag{10.1''}$$

Now we shall seek to transform x to z by means of a rotation matrix R, as we transformed s to z in Equations 10.22–10.26, using the orthogonal property of R:

$$x = Rz \text{ and } \dot{x} = R\dot{z} \tag{10.22''}$$

and

$$\tilde{x} = \tilde{z}\tilde{R} \text{ and } \tilde{\dot{x}} = \tilde{\dot{z}}\tilde{R}. \tag{10.24'}$$

Now, from Equation 10.1″, we have

$$\begin{aligned} L &= \tfrac{1}{2}\tilde{\dot{z}}\tilde{R}R\dot{z} - \tfrac{1}{2}\tilde{z}\tilde{R}URz \\ &= \tfrac{1}{2}\tilde{\dot{z}}\dot{z} - \tfrac{1}{2}\tilde{z}Kz, \end{aligned} \tag{10.3''}$$

where obviously

$$K = \tilde{R}UR = R^{-1}UR. \tag{10.34}$$

Then we shall have solved our problem if we can choose the rotation matrix R in such a way that K, the transformed U, is diagonal. In the process we shall have reduced the diagonal matrix M to the unit matrix I by having made our change of scale; the masses that previously appeared in M have now been absorbed into U and hence into K. These changes will not affect the calculated motion of the system. After the time-dependent z_i have been found in a given problem, they can be interpreted in terms of the original s_i via Equations 10.22′ and 10.32.

Equation 10.34 can be multiplied from the left by R, whence

$$UR = RK,$$

or (10.34′)

$$(UR)_{ij} = \sum_n U_{in}R_{nj} = (RK)_{ij} = \sum_n R_{in}K_{nj} = R_{ij}K_j,$$

where in the last step the diagonal form of K has been used, K_j being the diagonal element K_{jj}. In an N-dimensional space there are N^2 of these simultaneous linear equations for the unknown elements R_{ij} of the rotation matrix R that *diagonalizes* U (converts it to diagonal form).

In order to study these equations, let us look at the subset of N equations with a given fixed value of j. Let us temporarily suppress the subscript j. In doing this, we are confining our attention to one column of R and one of the diagonal elements of K. For the time being we can regard the R_n as being the elements of a column matrix or a vector, and regard K as being a single number. Then, upon writing out the equations, we find

$$\begin{aligned}
(U_{11} - K)R_1 + U_{12}R_2 + U_{13}R_3 + \dots + U_{1N}R_N &= 0 \\
U_{21}R_1 + (U_{22} - K)R_2 + U_{23}R_3 + \dots + U_{2N}R_N &= 0 \\
\vdots \qquad\qquad \vdots \qquad\qquad \vdots \\
U_{N1}R_1 + U_{N2}R_2 + \dots \qquad + (U_{NN} - K)R_N &= 0
\end{aligned} \tag{10.34''}$$

Here we have moved the K terms to the left and have omitted the index j that appears in Equation 10.34′. It is clear that the coefficients U_{in} and their arrangement in the equations are independent of the suppressed index j. Changing j would introduce a new element of K (a new number K) and would involve elements from a new column of R—but since we do not know either the elements of R or those of K, the new equations would confront us with exactly the same problem as do the old ones. If we can solve these equations for the R_n and in the process find out what K is, we should thus obtain enough information to determine all the K_j and all the R_{nj}.

We have N homogeneous equations for the N unknown R_n. Such a set has a solution only if the determinant of the coefficients equals zero.[2]

$$\begin{vmatrix} (U_{11} - K) & U_{12} & U_{13} & \cdots & U_{1N} \\ U_{21} & (U_{22} - K) & U_{23} & \cdots & U_{2N} \\ U_{31} & \cdot & \cdot & \cdots & U_{3N} \\ \cdot & \cdot & \cdot & \cdots & \cdot \\ \cdot & \cdot & \cdot & \cdots & \cdot \\ \cdot & \cdot & \cdot & \cdots & \cdot \\ U_{N1} & \cdot & \cdot & \cdots & (U_{NN} - K) \end{vmatrix} = 0 \qquad (10.35)$$

Since all the U_{mn} are fixed, we can secure the vanishing of this determinant (known as the *secular* determinant of the problem) only by making a suitable choice of the unknown K. Equation 10.35 (the *secular* equation) is an algebraic equation of Nth degree in the unknown K, so it has N roots. Each of these roots K_j can be substituted into the Equations 10.34″, permitting them to be solved for the R_n (the R_{nj} going with the particular K value K_j). Once K has been chosen as a root of the secular equation, Equations 10.34″ are compatible. They then have too many solutions, for because the equations are homogeneous in the R_n, any set of R_n that satisfy the equations can all be multiplied by an common multiplier and will still satisfy the equations. Thus, at most, Equations 10.34″ determine the unknown R_n up to a factor. This can be seen in another way, also. Having inserted one of the permissible values of K, let us divide all the equations by one of the unknown R_n and then move to the other side the coefficients from which that R_n has been removed. Then there are N *in*-homogeneous equations for the $N - 1$ unknown ratios of other R_n to the one that was used as a divisor. Any $N - 1$ of these may be soluble for the $N - 1$ ratios. If so, the omitted Nth equation will be satisfied by these solutions because the determinant of the original $N \times N$ array is zero. However, possibly every $(N - 1) \times (N - 1)$ determinant in the original $N \times N$ array is zero. In this case, one cannot solve $N - 1$ equations for the $N - 1$ ratios of elements of R; the original equations do not even determine a vector R up to a factor, but may possibly tell what plane (instead of what line) the vector lies in, or what three-space, etc.

We have seen that it is always possible to choose at least one component of R arbitrarily, for Equations 10.34″ determine R at most only up to a factor. If they determine it less fully than that, more of its components are arbitrary. The degree of arbitrariness depends on the multiplicities of the roots of the secular equation. If all the roots are distinct, they are N different numbers, each of which determines a vector R up to a factor. If, instead, one root is double, that value of K determines a plane in which two of the N vectors must lie; a triple root determines a three-space occupied by three of the vectors, etc.

In the equation

$$UR = KR, \tag{10.34''}$$

where U is a square matrix, R is a column matrix, and K is a number, K is spoken of as an *eigenvalue* (also *characteristic* value) of U, and R is the *eigenvector* (*characteristic* vector) of U belonging to the eigenvalue K. In our search for the square matrix R that diagonalizes U we have been looking for one column at a time. Each column has had to satisfy Equation 10.34″, so has been an eigenvector. The diagonal elements of the matrix K (the diagonalized form of U) are the eigenvalues of U.

We can easily establish some important properties of the eigenvectors and eigenvalues of a real symmetric matrix U. Let us write two eigenvalue equations for U, involving two eigenvalues K and K' (which may be the same number), and two eigenvectors R and R' (which may be the same vector). Then we take the transpose complex conjugate of one of the equations:

$$\begin{aligned} UR = KR;\ \tilde{R}^*\tilde{U}^* = \tilde{R}^*\, U = K^*\, \tilde{R}^* \\ UR' = K'R'. \end{aligned} \tag{10.36}$$

Now we multiply the upper, transposed equation from the right by R' and the lower from the left by $\tilde{R}^*$ and subtract one from the other:

$$\begin{aligned} \tilde{R}^*UR' &= K^*\, \tilde{R}^*\, R'; \\ \tilde{R}^*UR' &= K'\, \tilde{R}^*\, R'; \\ 0 &= (K^* - K')(\tilde{R}^*R'). \end{aligned} \tag{10.37}$$

The equality of the two left sides, whereby the last equation has zero on the left, is a consequence of U being real and symmetric. Now in the last equation a product of two factors equals zero, so at least one of the factors must be zero, the factors being numbers rather than more elaborate objects. First, then, let us choose K and K' to be the same number and R and R' to be the same vector. Then the product $\tilde{R}^*R$ is positive-definite, being a sum of squares of absolute values of the components of R; it can be zero only in the trivial case in which R is a zero vector. Thus we learn that K equals its own complex conjugate, or it is real; the roots of the secular equation are all real. So a real symmetric matrix has all real eigenvalues.

If the eigenvalues K are all real, the coefficients in the Equations 10.34″ are all real, and so all the components R_n can be chosen real. Thus we can now drop the complex conjugate marks in Equations 10.37. Now let us assume that $K \neq K'$ in the last of these equations. If $K \neq K'$, $\tilde{R}\,R' = 0$: eigenvectors of a real symmetric matrix belonging to different eigenvalues are orthogonal.

If $K = K'$, this root may be multiple; that is, there may be two or more independent vectors R belonging to this eigenvalue. Such an eigenvalue is spoken of as *degenerate*—k-fold degenerate when k eigenvalues of U coincide. Equation 10.37 does not tell us anything about the product of two eigenvectors belonging to the same eigenvalue. However, we can learn something about this case by noting that, if R and R' are two eigenvectors of U belonging to the common eigenvalue K,

$$
\begin{aligned}
UR &= KR, \\
UR' &= KR'.
\end{aligned}
\tag{10.38}
$$

it follows that any linear combination of R and R' is also an eigenvector of U belonging to the same eigenvalue K,

$$
U(aR + bR') = K(aR + bR'). \tag{10.39}
$$

Thus, given two (or more) linearly independent eigenvectors of U belonging to the same eigenvalue, we can construct an equal number of orthogonal eigenvectors belonging to that eigenvalue by making that number of linear combinations of the original set and then requiring the linear combinations to be mutually orthogonal. Thus, any two eigenvectors of U either will be orthogonal by virtue of belonging to different eigenvalues, or can be chosen orthogonal when they belong to a common eigenvalue. We can, then, assume that they are all mutually orthogonal.

Furthermore, since all eigenvectors are indeterminate to at least the extent of a possible change of scale, all can be chosen to have unit length:

$$\tilde{R}\,R = 1. \tag{10.40}$$

Thus the eigenvectors of U form, or can be chosen to form, an orthonormal set like the base vectors **e** and **f** of Equations 10.20–10.30. Reinstating the label j for the columns of the square transformation matrix R, we have

$$\tilde{R}_j\,R_i = \sum_n R_{nj}\,R_{ni} = \delta_{ij}, \tag{10.41}$$

or the square matrix R is orthogonal; it produces a rotation of axes. So we have shown that a real symmetric matrix can be diagonalized by a (properly chosen) rotation. We have shown that, as suggested after Equation 10.34, the rotation matrix R can be chosen so as to diagonalize U, and we have seen how to find an R that will do the job—make it up out of orthonormal eigenvectors of U, placed in the respective columns of R. When the right R has been found, it transforms U into the diagonal matrix K:

$$\tilde{R}\,UR = R^{-1}\,UR = K, \tag{10.42}$$

the diagonal elements of which are the eigenvalues of U, the K_j.

Thus we have solved the problem of small vibrations in the special case in which the kinetic-energy matrix T is diagonal at the outset. The general case, in which T is real and symmetric but not diagonal, can now be handled easily. Writing

$$L = \tfrac{1}{2}\tilde{\dot{s}}\ T\ \dot{s}\ -\ \tfrac{1}{2}\tilde{s}\ V\ s, \tag{10.1'}$$

we first seek a rotation S that diagonalizes T:

$$\tilde{S}\ T\ S = S^{-1}\ T\ S = M, \text{ a diagonal matrix.} \tag{10.42}$$

T being real and symmetric, it can be diagonalized as was U in the argument just completed. In the process, V gets changed, also, to

$$V' = \tilde{S}\ V\ S = S^{-1}V\ S. \tag{10.43}$$

S being orthogonal, one can easily prove that, if V is real and symmetric, then V' is also:

$$V'_{ij} = \sum_{nm} \tilde{S}_{in} V_{nm} S_{mj} = \sum_{nm} S_{ni} V_{nm} S_{mj} = \sum_{nm} S_{ni} V_{mn} S_{mj} =$$
$$\sum_{nm} \tilde{S}_{jm} V_{mn} S_{ni} = V'_{ji}, \tag{10.44}$$

where the symmetry of V was used in the third step. V' is symmetric, and has to be real because all its ingredients are. So, if

$$Sy = s, \tag{10.45}$$

we have L in the form

$$L = \tfrac{1}{2}\tilde{\dot{y}}\ M\ \dot{y}\ -\ \tfrac{1}{2}\tilde{y}\ V'\ y, \tag{10.46}$$

where M is diagonal and V' is real and symmetric. Now we are back at the point at which we began our previous argument, and by the change of scale involving the diagonal elements of M, Equation 10.32, we can bring L into the form of Equation 10.1″). Then, as previously, we have the unit matrix I in the kinetic energy term, and we know that the rotation R that will diagonalize U will not spoil the convenient form of T. Then, applying R, we have both parts of L reduced to sums of squares, as in Equation 10.3. Starting with square matrices T and V, and column matrix s, we go ultimately to diagonal square matrices I (not written) and K, and column matrix z, by the sequences of steps

$$s = SM^{-1/2}Rz, \tag{10.47}$$

and

$$K = \tilde{R}\, M^{-1/2}\, \tilde{S}\, V\, S\, M^{-1/2}\, R,$$

where $M^{-1/2}$ is a diagonal square matrix with elements respectively equal to the reciprocal square roots of those of the diagonal matrix M, that is, of $\tilde{S}\ V\ S$.[3]

When the transformation to z and K is complete, the Lagrangian is

$$L = \tfrac{1}{2}\tilde{\dot{z}}\ \dot{z}\ - \tfrac{1}{2}\tilde{z}\ K\ z, \tag{10.48}$$

where K is diagonal, with elements K_j, and so the equations of motion are

$$\ddot{z}_j + K_j z_j = 0, \tag{10.4'}$$

an even simpler form than Equation 10.4 because the masses have been absorbed into the denominators of the $\tilde{K}_j$. Thus, z_j oscillates with frequency $\omega_j = K_j^{1/2}$.

Although one usually calculates normal coordinates and frequencies in order to study a system's undriven oscillations about equilibrium, sometimes it is desirable to use them in determining responses of systems to known applied forces. Let us then insert such forces in Equation 9.26:

$$\sum_n T_{mn}\, \ddot{s}_n + \sum_n V_{mn} s_n = F_m(t),$$

or (10.49)

$$T\ddot{s} + Vs = F(t),$$

where the proper use of F_n has been disinterred from Equation 9.6 and the second form above is simply a matrix abbreviation of the first. Now, with the help of Equations 10.42 and 10.47, we can write

$$TSM^{-1/2}R\ \ddot{z} + VSM^{-1/2}Rz = F(t),$$

whence (10.50)

$$\ddot{z} + Kz = \tilde{R}M^{-1/2}\tilde{S}F(t),$$

where the second equation follows from the first by means of multiplication from the left by $\tilde{R}\, M^{-1/2}\, \tilde{S}$. Thus, K and M (and hence $M^{-1/2}$) being diagonal, we can write in terms of components

$$\ddot{z}_j + K_j z_j = \sum_{nm} R_{nj}\, S_{mn}\, F_m(t)/\sqrt{M_n} \equiv f_j(t). \tag{10.51}$$

Here the M_n are the eigenvalues of the T matrix, as indicated in Equation 10.42, and the definitions of the $f_j(t)$ are obvious. Thus each normal coordinate z_j is subject to a driving force that is an appropriate linear combination of the $F_m(t)$ which were assumed to be acting on the original coordinates s_n in Equation 10.49. Once we have Equation 10.51, we simply have to calculate the responses of a set of noninteracting oscillators to a set of known forces. Thus the circuit problem of Chapter 7 has been dissected into as many independent one-loop problems as the number of loops in the original circuit. However, the present treatment lacks generality, because the system has been assumed to have no damping; the circuit contains no resistances. As was mentioned in Chapter 7, such a system has all the poles of $h(\omega)$ on the real axis of ω. If we wish to evaluate Fourier integrals for such a system, we have to move the poles a small distance above the real axis. Here, each normal-mode oscillator responds independently. Each has two first-order poles (or if $K_j = 0$, one second-order pole) on the real axis. Each of these poles has to be interpreted as being slightly above the real axis.

It is possible, but not very useful, to apply the foregoing procedures to dissipative systems. Equations 7.1 can be written in matrix notation as

$$F = ZI. \tag{10.52}$$

Solving the equations amounts to finding the reciprocal of the symmetric (complex) matrix Z:

$$I = Z^{-1}\, F. \tag{10.53}$$

We can, instead, diagonalize Z by finding its eigenvalues and eigenvectors, and making a transformation matrix out of the latter, with

$$Z' = \tilde{R}ZR = \text{diagonal}, \tag{10.54}$$

$$\tilde{R} = R^{-1}.$$

Then

$$RF = Z'RI,$$

or (10.55)

$$(RF)_j = Z'_j(RI)_j,$$

Z' being diagonal by definition. Although Z is a complex matrix, all the theorems about its eigenvalues and eigenvectors and diagonal form still hold except the one about eigenvalues' being real. So Equation 10.55 contains a one-to-one linear frequency-dependent relationship between linear combinations of the applied voltages and the same combinations of the currents—as if the circuit had been dissected into independent loops, each containing a complex impedance. The procedure is difficult to interpret, for the currents involved in the equations are steady-state currents and so cannot directly give information about undriven behavior of the system. Furthermore, the matrix Z is frequency-dependent, so also are its eigenvalues and eigenvectors. Also, the linear combinations of voltages and currents, given by the rotation matrix R, have frequency-dependent coefficients. Thus the interpretation of Fourier analysis applied to the separate uncoupled circuits is very difficult, and the procedure does not seem to be well suited to the solving of problems in which dissipative effects play a part.

NOTES

1. For a full and careful treatment of matrix algebra, the reader is referred to books on higher algebra; for example, Bôcher, M. *In-*

troduction to Higher Algebra. New York: Macmillan, 1924.

2. The properties of systems of linear algebraic equations are discussed in Arfken, G. *Mathematical Methods for Physicists*. 2d ed. New York: Academic Press, 1970. Bôcher, *Introduction to Higher Algebra*. Courant, R., and D. Hilbert. *Methods of Mathematical Physics*. New York; Interscience, 1953; also Margenau, H. and G. M. Murphy. *The Mathematics of Physics & Chemistry*. New York: Van Nostrand, 1943.
3. It is customary to discuss the normal-coordinate transformation as a single process rather than as a sequence of three operations. (See, for example, Goldstein, H. *Classical Mechanics*. 2nd ed. Reading, MA: Addison Wesley, 1980.

PROBLEMS

10.1 Given column matrices r and s and a square matrix W, such that $s = Wr$, show that there is not a unique W that satisfies this equation with given r, s. How many pairs of column matrices thus related by W are required to determine W uniquely?

10.2 Let A be a square matrix and I the identity of the same size.
(a) Show that if $XA = I$, then $AX = I$;
(b) Show that the inverse of A, if there is one, is unique.

10.3 If A and B are similar square matrices, under what circumstances is

$$(A + B)(A - B) = A^2 - B^2?$$

10.4 Show that the product of two orthogonal matrices is orthogonal.

10.5 Show that the following properties of a square matrix are invariant under an orthogonal transformation:
(a) the trace (i.e., the sum of diagonal elements),
(b) the determinant,
(c) the symmetry or antisymmetry.

10.6 If A is a real symmetric matrix, show that
(a) its trace equals the sum of its eigenvalues;
(b) its determinant equals plus or minus the product of its eigenvalues.

10.7 A certain square matrix, not the unit matrix, equals its own square. Such a matrix is known as a projection operator. Show that it is singular, has eigenvalues 0 and 1, and equals each of its powers. What does it do to the vectors in a vector space?

11

Some Examples of Normal Coordinates

In this chapter we shall apply the methods of Chapter 10 to specific physical systems. Let us take as our first example two masses, m_1 and m_2, sliding without friction on two parallel rods separated by distance D. The masses are connected by a light spring of spring constant k_0 and unstretched length $S_0 < D$. Letting their positions on their rods be s_1 and s_2, we use the analysis of Equations 1.10–1.15 to approximate their potential energy. Thus we have

$$L = \tfrac{1}{2}(m_1\dot{s}_1^2 + m_2\dot{s}_2^2) - \tfrac{1}{2}k(s_2 - s_1)^2, \tag{11.1}$$

where

$$k = k_0(1 - S_0/D), \tag{11.2}$$

and the constant U_0 of Chapter 1 has been dropped. So the matrices T and V are

$$T = \begin{pmatrix} m_1 & 0 \\ 0 & m_2 \end{pmatrix}; \quad V = \begin{pmatrix} k & -k \\ -k & k \end{pmatrix}; \tag{11.3}$$

where T is obviously diagonal already (but not yet a multiple of I), and V has been defined to be symmetric. Now letting $s_i = x_i m_i^{-1/2}$, we have

$$L = \tfrac{1}{2}\dot{\tilde{x}}\,\dot{x} - \tfrac{1}{2}\,\tilde{x}\,U\,x, \tag{11.4}$$

where

$$U = \begin{pmatrix} k/m_1 & -k/(m_1 m_2)^{1/2} \\ -k/(m_1 m_2)^{1/2} & k/m_2 \end{pmatrix}, \tag{11.5}$$

which is the matrix whose eigenvalues and eigenvectors we now need to find. $UR = KR$:

$$\begin{aligned} (k/m_1 - K)R_1 - [k/(m_1 m_2)^{1/2}]R_2 &= 0; \\ -[k/(m_1 m_2)^{1/2}]R_1 + (k/m_2 - K)R_2 &= 0. \end{aligned} \tag{11.6}$$

The secular equation is

$$K^2 - Kk(1/m_1 + 1/m_2) + k^2/m_1 m_2 - k^2/m_1 m_2 = 0, \tag{11.7}$$

and its roots are

$$\begin{aligned} K_1 &= 0; \\ K_2 &= k(1/m_1 + 1/m_2) = k/\mu; \end{aligned} \tag{11.8}$$

where μ is the reduced mass of the two objects. Now the first column of the transformation matrix R corresponds to $K = 0$, and the second column corresponds to $K = k/\mu$.

Thus the first equation tells us that

$$R_{11} = (m_1/m_2)^{1/2}\,R_{21}, \tag{11.9}$$

and the second equation, as is to be expected, contains the same information. The most that we can hope for is that the equations will determine a column of R up to a factor, and Equation 11.9 already does that. Thus, if Equation 11.9 did not identically satisfy the second Equation 11.6, it would have to contradict it—a sign that K had been wrongly calculated. Having thus determined R_{11} and R_{21} up to a factor, we now impose, as a separate requirement, the condition that the first column of R is a *unit* vector:

$$R_{11}{}^2 + R_{21}{}^2 = 1. \tag{11.10}$$

One solution (determined up to a sign) of Equation 11.9 and 11.10 is

$$\begin{aligned} R_{11} &= [m_1/(m_1 + m_2)]^{1/2}; \\ R_{21} &= [m_2/(m_1 + m_2)]^{1/2}. \end{aligned} \tag{11.11}$$

The second column of R is calculated by similar methods. The first of Equations 11.6, with $K = k/\mu = k(1/m_1 + 1/m_2)$, yields

$$R_{12} = -(m_2/m_1)^{1/2} R_{22}, \tag{11.12}$$

and again the correctness of K is checked by the second Equation 11.6 giving exactly the same information. Again the *normalization* of the second column of R (sum of squares = 1) provides additional information that determines the two elements up to a sign. They are

$$\begin{aligned} R_{12} &= -\{m_2/(m_1 + m_2)\}^{1/2}; \\ R_{22} &= \{m_1/(m_1 + m_2)\}^{1/2}. \end{aligned} \tag{11.13}$$

Thus the entire rotation matrix R is

$$R = (m_1 + m_2)^{-1/2} \begin{pmatrix} m_1^{1/2} & -m_2^{1/2} \\ m_2^{1/2} & m_1^{1/2} \end{pmatrix}. \tag{11.14}$$

This matrix is obviously orthogonal; it has orthogonal rows and orthogonal columns, and (counting the factor in front) each row and each column is normalized, i.e., is a unit vector. The determinant of the matrix (including the factor in front) is +1. This simple configuration space can be drawn in two dimensions and the rotation exhibited (see Figure 11.1).

These two dimensions should not be confused with directions in space; x_1 and x_2 are orthogonal directions in configuration space, but correspond (with changed scales) to parallel displacements of two masses in real space. The angle of rotation, ϕ, in Figure 11.1 appears in transformation equation(s) $x = Rz$ as follows:

$$
\begin{aligned}
x_1 &= z_1 \cos\phi - z_2 \sin\phi; \\
x_2 &= z_1 \sin\phi + z_2 \cos\phi;
\end{aligned}
\tag{11.15}
$$

where

$$
\cos\phi = [m_1/(m_1 + m_2)]^{1/2}; \tag{11.16}
$$

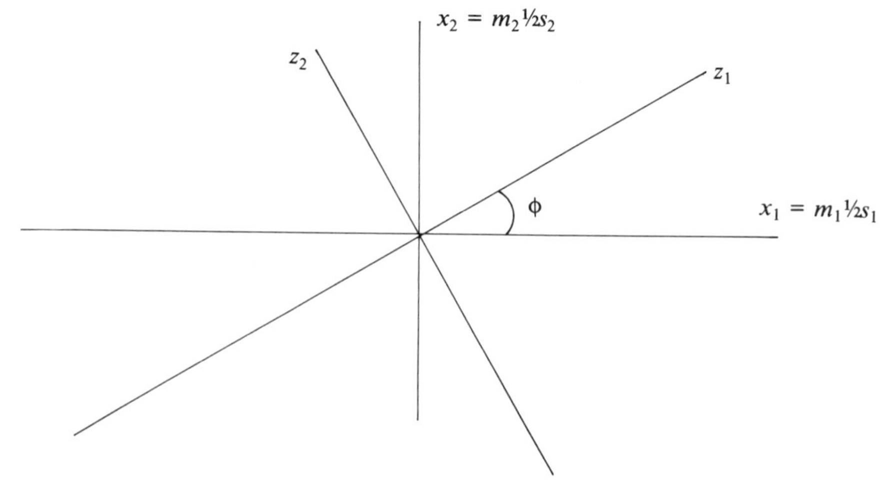

Figure 11.1. The rotation in configuration space corresponding to the normal-coordinate transformation of Equations 11.15–11.18.

$$\sin\phi = [m_2/(m_1 + m_2)]^{1/2}.$$

Then the original coordinates s_i are expressed in terms of the normal coordinates z_i as:

$$s_1 = x_1/m_1^{1/2} = z_1(m_1 + m_2)^{-1/2} - z_2[m_2/m_1(m_1 + m_2)]^{1/2}; \tag{11.17}$$

$$s_2 = x_2/m_1^{1/2} = z_1(m_1 + m_2)^{-1/2} + z_2[m_1/m_2(m_1 + m_2)]^{1/2}.$$

The inverse relations are $z = \tilde{R}\, x$:

$$z_1 = x_1 \cos\phi + x_2 \sin\phi = s_1 m_1/(m_1 + m_2)^{1/2} + s_2 m_2/(m_1 + m_2)^{1/2}; \tag{11.18}$$

$$z_2 = -x_1 \sin\phi + x_2 \cos\phi = (-s_1 + s_2)\,[m_1 m_2/(m_1 + m_2)]^{1/2}.$$

Thus we see that the normal coordinate z_1, the one with $K = 0$, or zero frequency (see the end of Chapter 10), is proportional to the coordinate of the center of mass, as it had to be; any isolated system—as ours is, in one dimension—has a center of mass that moves with constant velocity, which is the limiting case of simple harmonic motion as frequency goes to zero. The normal coordinate z_2, with frequency $K_2^{1/2} = (k/\mu)^{1/2}$, is proportional to the relative coordinate of the two masses, $s_2 - s_1$—again, as it had to be; the motion of an isolated two-body system can always be separated into unaccelerated center-of-mass motion and relative motion and the latter is equivalent to that of a single body of mass μ subject to the force that acts between the two bodies. Thus we could have solved the present problem by means of general principles, without using matrices, eigenvalues, etc. This same fact, however, makes the problem a good first example of the normal-coordinate calculation, for there are well-known independent checks on the results.

As defined earlier, the normal modes of motion are the ways in which the system can move in SHM, that is, the motions that involve only one normal coordinate at a time. The general solutions of the z equations of motion. Equation 10.4′, are, in our case

$$z_1 = a + bt;$$

$$z_2 = c\cos(\omega_2 t + \theta); \tag{11.19}$$

$$\omega_2 = (k/\mu)^{1/2}.$$

The general motions of the s_i are then obtained through substitution of these solutions into Equations 11.17. Each of the two normal modes is obtained when only one z is nonzero. Thus the normal mode associated with z_1 is

$$s_1 = s_2 = (a + bt)(m_1 + m_2)^{-1/2} = A + Bt, \tag{11.20}$$

i.e., motion of the center of mass with constant velocity, and no internal motion. The second normal mode is

$$s_1 = -(m_2/m_1)^{1/2}(m_1 + m_2)^{-1/2}\, c\cos(\omega_2 t + \theta);$$

$$\tag{11.21}$$

$$s_2 = (m_1/m_2)^{1/2}(m_1 + m_2)^{-1/2}\, c\cos(\omega_2 t + \theta).$$

Here $m_1 s_1 + m_2 s_2 = 0$; the center of mass does not move. All the motion is relative motion of the two masses, with the normal frequency ω_2. Figure 11.2 shows these motions schematically.

In order to fit a solution to given initial values $s_1(0), s_2(0), \dot{s}_1(0), \dot{s}_2(0)$, we can use Equations 11.18 and 11.19 to get

$$z_1(0) = a = s_1(0)m_1/(m_1 + m_2)^{1/2} + s_2(0)m_2/(m_1 + m_2)^{1/2};$$

$$z_2(0) = c\cos\theta = [s_2(0) - s_1(0)]\, m_1 m_2 (m_1 + m_2)^{-1/2};$$

$$\tag{11.22}$$

$$\dot{z}_1(0) = b = \dot{s}_1(0)m_1/(m_1 + m_2)^{1/2} + \dot{s}_2(0)m_2/(m_1 + m_2)^{1/2};$$

$$\dot{z}_2(0) = -c(k/\mu)^{1/2}\sin\theta$$

$$= [\dot{s}_2(0) - \dot{s}_1(0)]m_1 m_2 (m_1 + m_2)^{-1/2}.$$

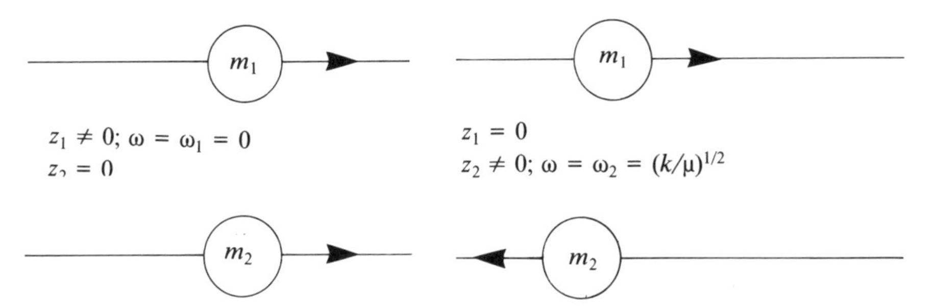

Figure 11.2. Normal modes of motion described in Equation 11.19.

From these equations we should be able to solve for a, b, c, θ, which can then be used in Equations 11.19 and 11.17 to give $s_1(t)$ and $s_2(t)$.

As our second example, let us again consider two masses sliding without friction on parallel rods separated by distance D and connected by a spring of unstretched length $S_0 < D$. But now let us take the masses to be equal, and let us assume that each is connected to a fixed point by another spring, again of unstretched length S_0. Let the fixed points be opposite each other, each a distance D from the rod on which slides the mass connected to that point by a spring. Let the effective spring constant $k_0(1 - S_0/D)$ of the spring between the masses be now called k', assumed *less* than the corresponding common quantity k for the other springs. The system is depicted in Figure 11.3, with its displacements from equilibrium, s_i, exaggerated in comparison with D.

Now the Lagrangian is, for small displacements ($s_i \ll D$),

$$L = \tfrac{1}{2}m(\dot{s}_1^2 + \dot{s}_2^2) - \tfrac{1}{2}k(s_1^2 + s_2^2) - \tfrac{1}{2}k'(s_2 - s_1)^2, \quad (11.23)$$

where we have used the treatment in Chapter 1 again, have again discarded the constant terms in V, and have replaced the original spring constants by the equivalent values k and k' as these values are calculated in Chapter 1.

This L can be written as

$$L = \tfrac{1}{2}m\,\tilde{\dot{s}}\,\dot{s} - \tfrac{1}{2}\,\tilde{s}\,V\,s, \quad (11.24)$$

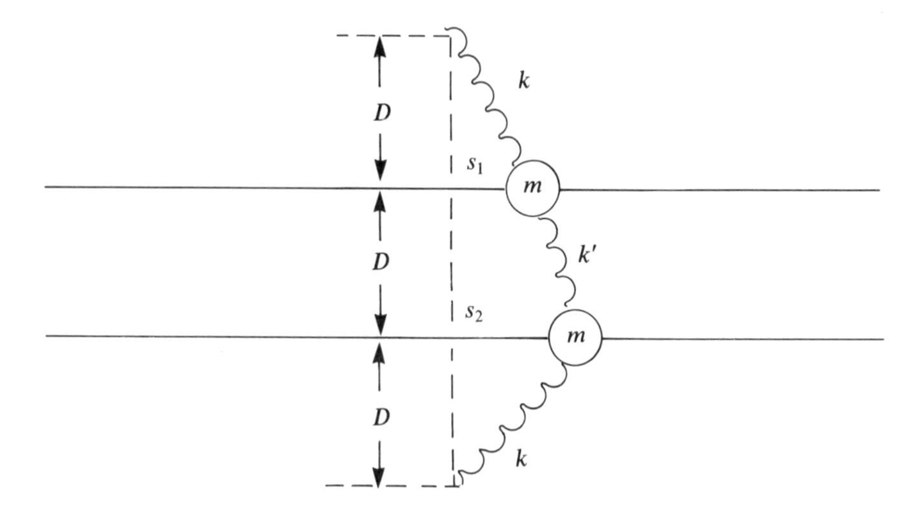

Figure 11.3. The system with Lagrangian 11.23.

where

$$V = \begin{pmatrix} k + k' & -k' \\ -k' & k + k' \end{pmatrix}. \tag{11.25}$$

Here the matrix T is already a multiple of I, so it will be invariant under rotation; there is no need to render it invariant by dividing by square roots of masses. We can diagonalize V as it stands. So the secular equation is

$$\begin{vmatrix} k + k' - K & -k' \\ -k' & k + k' - K \end{vmatrix} = K^2 - 2K(k + k') + k^2 + 2kk' = 0, \tag{11.26}$$

and has roots

$$\begin{aligned} K_1 &= k; \\ K_2 &= k + 2k'. \end{aligned} \tag{11.27}$$

Now the first eigenvector, going with K_1, satisfies

$$k'R_{11} - k'R_{21} = 0;$$
$$-k'R_{11} + k'R_{21} = 0. \tag{11.28}$$

As expected, these equationas have the same content, that $R_{11} = R_{21}$. Normalizing the vector, we have

$$R_{11} = R_{21} = 2^{-1/2}. \tag{11.29}$$

The second eigenvector satisfies

$$-k'R_{12} - k'R_{22} = 0;$$
$$-k'R_{12} - k'R_{22} = 0. \tag{11.30}$$

So the normalized vector has components

$$-R_{12} = R_{22} = 2^{-1/2}. \tag{11.31}$$

Thus the rotation matrix R is

$$R = \sqrt{\tfrac{1}{2}}\begin{pmatrix} 1 & -1 \\ 1 & 1 \end{pmatrix}, \tag{11.32}$$

or the angle ϕ analogous to that in Equation 11.15 and 11.18 is now 45°. Again R is obviously orthogonal. The equation $s = Rz$ is

$$s_1 = (z_1 - z_2)/\sqrt{2};$$
$$s_2 = (z_1 + z_2)/\sqrt{2}. \tag{11.33}$$

The inverse transformation, $z = R^{-1} s = \tilde{R} s$ is

$$z_1 = (s_1 + s_2)/\sqrt{2};$$
$$z_2 = (s_2 - s_1)/\sqrt{2}. \tag{11.34}$$

The Lagrangian, reduced to a sum of squares, leads to the equations of motion

$$m\ddot{z}_1 = -K_1 z_1;$$
$$m\ddot{z}_2 = -K_2 z_2. \qquad (11.35)$$

These equations have the general solutions

$$z_1 = A_1 \cos(\omega_1 t + \theta_1);$$
$$z_2 = A_2 \cos(\omega_2 t + \theta_2). \qquad (11.36)$$

Here $\omega_1 = (K_1/m)^{1/2}$ and $\omega_2 = (K_2/m)^{1/2}$. Initial conditions can be applied by means of Equations 11.34 and 11.36:

$$z_1(0) = A_1 \cos\theta_1 = [s_1(0) + s_2(0)]/\sqrt{2};$$
$$z_2(0) = A_2 \cos\theta_2 = [s_2(0) - s_1(0)]/\sqrt{2};$$
$$\dot{z}_1(0) = -\omega_1 A_1 \sin\theta_1 = [\dot{s}_1(0) + \dot{s}_2(0)]/\sqrt{2};$$
$$\dot{z}_2(0) = -\omega_2 A_2 \sin\theta_2 = [\dot{s}_2(0) - \dot{s}_1(0)]/\sqrt{2}. \qquad (11.37)$$

Such information can be used in Equations 11.33 and 11.36 to determine the $s_i(t)$. For instance, suppose that the system is set in motion by mass #1 being released from rest at $s_1 = A$, while mass #2 is at rest at the origin. Then $s_1(0) = A$, and the other initial quantities = 0.

Now, by Equations 11.36,

$$\sin\theta_1 = \sin\theta_2 = 0;\ \cos\theta_1 = 1;\ \cos\theta_2 = -1;$$
$$A_1 = A_2 = A/\sqrt{2}. \qquad (11.38)$$

These values in Equations 11.36 yield

$$z_1 = 2^{-1/2} A \cos\omega_1 t; \qquad (11.39)$$

$$z_2 = -2^{-1/2}A\cos\omega_2 t.$$

Then, with the help of Equation 11.33 we find that

$$s_1 = \tfrac{1}{2}A\,(\cos\omega_1 t + \cos\omega_2 t) = A\cos\tfrac{1}{2}(\omega_1 + \omega_2)t\,\cos\tfrac{1}{2}(\omega_2 - \omega_1)t; \tag{11.40}$$

$$s_2 = \tfrac{1}{2}A\,(\cos\omega_1 t - \cos\omega_2 t) = A\sin\tfrac{1}{2}(\omega_1 + \omega_2)t\,\sin\tfrac{1}{2}(\omega_2 - \omega_1)t.$$

Here the last step on each line results from a trigonometric identity. Now, recalling that we have assumed $k > k'$ and have calculated $\omega_1 = (k/m)^{1/2}$ and $\omega_2 = [(k + 2k')/m]^{1/2}$, we can approximate Equations 11.40 for the extreme case $k \gg k'$:

$$\omega_2 = \omega_1(1 + 2k'/k)^{1/2} \cong \omega_1(1 + k'/k);$$

$$\omega_2 + \omega_1 = \omega_1(2 + k'/k) \cong 2\omega_1; \tag{11.41}$$

$$\omega_2 - \omega_1 \cong \omega_1\,k'/k = k'/(km)^{1/2}.$$

Thus Equation 11.40 shows that both masses will oscillate with frequency $\tfrac{1}{2}(\omega_2 + \omega_1) \cong \omega_1$, with each oscillation having a slowly changing amplitude. Initially only mass #1 is moving, for it is the one that was released from rest, but after $k/2k'$ cycles of this motion the cosine function of $\tfrac{1}{2}(\omega_2 - \omega_1)t \cong (k'/2k)\omega_1 t$ will have gone to zero, and the sine of the same quantity will be maximum. At this time mass #1 will have come to rest, and mass #2 will be oscillating with the amplitude with which mass #1 began. An equal time later still mass #1 will again be moving maximally, and mass #2 will be at rest. This alternation of motion will continue, with one cycle of the alternation occupying k/k' cycles of the more rapid oscillation. This situation, sometimes demonstrated by means of weakly coupled pendulums, is obviously *not* a case of simple harmonic motion; both normal coordinates z_1 and z_2 are participating.

However, the normal modes of motion are readily derived and depicted. Letting only z_1 be nonzero, we have, from Equation

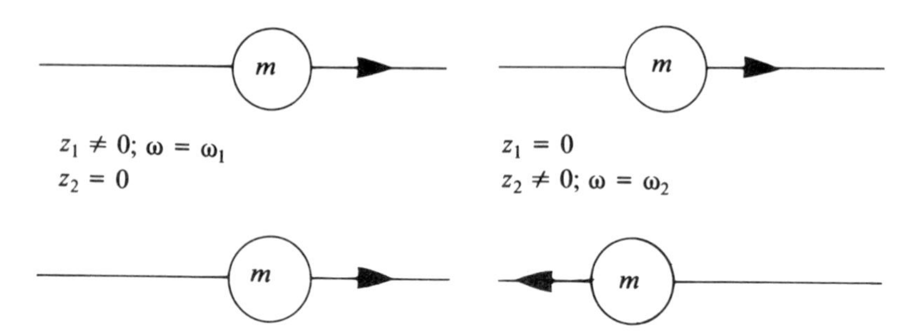

Figure 11.4. Normal modes of motion for the Lagrangian 11.23.

11.33, $s_1 = s_2$; both have frequency $\omega_1 = (k/m)^{1/2}$ and are moving in SHM. When only z_2 is nonzero, $s_2 = -s_1$, and both are moving in SHM. These motions are sketched in Figure 11.4.

Our third example is a system of three equal masses sliding without friction on three parallel rods and connected in pairs by three equal springs; the rods intersect a plane perpendicular to them at the vertices of an equilateral triangle; their spacing is D, greater than the unstretched lengths of the three springs. Thus, if k is the effective spring constant,

$$\begin{aligned} L &= \tfrac{1}{2}m \sum_i \dot{s}_i^2 - \tfrac{1}{2}k[(s_1 - s_2)^2 + (s_2 - s_3)^2 + (s_3 - s_1)^2] \\ &= \tfrac{1}{2}m\, \tilde{\dot{s}}\, \dot{s} - \tfrac{1}{2}\, \tilde{s}\, V\, s, \end{aligned} \tag{11.42}$$

where

$$V = k\begin{pmatrix} 2 & -1 & -1 \\ -1 & 2 & -1 \\ -1 & -1 & 2 \end{pmatrix}. \tag{11.43}$$

Since the matrix in the kinetic-energy term is a multiple of the identity, we need only to find an orthogonal matrix that diagonalizes V. Thus we set $VR = KR$, or, letting $K = kw$, we write

$$\begin{aligned} (2 - w)R_1 - R_2 - R_3 &= 0; \\ -R_1 + (2 - w)\, R_2 - R_3 &= 0; \\ -R_1 - R_2 + (2 - w)R_3 &= 0. \end{aligned} \tag{11.44}$$

The secular equation is

$$\begin{vmatrix} (2-w) & -1 & -1 \\ -1 & (2-w) & -1 \\ -1 & -1 & (2-w) \end{vmatrix} = -w^3 + 6w^2 - 9w = 0$$
$$= -w(w-3)^2. \quad (11.45)$$

One eigenvalue is zero; its eigenvector is given by

$$\begin{aligned} 2R_1 \; -R_2 \; -R_3 &= 0; \\ -R_1 \; +2R_2 \; -R_3 &= 0; \\ -R_1 \; -R_2 \; +2R_3 &= 0. \end{aligned} \quad (11.46)$$

The first two of these equations yield $R_1 = R_2 = R_3$, a solution that expectedly satisfies the third equation. Normalization of the vector then gives the value $3^{-1/2}$ for each component.

The other eigenvalue, $w = 3$ or $K = 3k$, is doubly degenerate; the root is double. Thus we expect to find two independent eigenvectors belonging to this eigenvalue. In the equations, it yields

$$-R_1 - R_2 - R_3 = 0 \quad (11.47)$$

three times; the three equations are all the same. Thus this vector is restricted only to a plane in the three-space (instead of a line as the first vector is). Any two independent vectors in that plane can be taken to be the two desired eigenvectors, but in order to get an orthogonal transformation matrix R we must choose the two vectors to be orthogonal and normalized.

We can arbitrarily choose $R_3 = 0$ in one of them, whence $R_2 = -R_1 = 2^{-1/2}$. Then the third vector can be chosen as $R_1 = R_2 = -6^{-1/2}, R_3 = 2/6^{1/2}$; thus it has unit length and is orthogonal to the second vector. Note that the second and third vectors are orthogonal to the first, as they had to be.

So the rotation matrix is

$$R = \begin{pmatrix} 1/\sqrt{3} & -1/\sqrt{2} & -1/\sqrt{6} \\ 1/\sqrt{3} & 1/\sqrt{2} & -1/\sqrt{6} \\ 1/\sqrt{3} & 0 & 2/\sqrt{6} \end{pmatrix}. \quad (11.48)$$

Thus $s = Rz$ gives

$$
\begin{aligned}
s_1 &= z_1/\sqrt{3} - z_2/\sqrt{2} - z_3/\sqrt{6}; \\
s_2 &= z_1/\sqrt{3} + z_2/\sqrt{2} - z_3/\sqrt{6}; \qquad (11.49) \\
s_3 &= z_1/\sqrt{3} \qquad\quad + 2z_3/\sqrt{6}.
\end{aligned}
$$

The inverse relationship $z = \tilde{R}s$ yields

$$
\begin{aligned}
z_1 &= (s_1 + s_2 + s_3)/\sqrt{3}; \\
z_2 &= (s_2 - s_1)/\sqrt{2}; \qquad (11.50) \\
z_3 &= (2s_3 - s_2 - s_1)/\sqrt{6}.
\end{aligned}
$$

Thus z_1 moves with zero frequency, i.e., as a free particle; it is proportional to the center-of-mass coordinate. The other two normal coordinates, which are somewhat arbitrarily defined, have the common frequency $(3k/m)^{1/2}$. Figure 11.5 depicts the three normal modes, as here defined.

The motions of z_2 and z_3, with frequency $(3k/m)^{1/2}$, could be any two independent motions that do not move the center of mass. The solution of initial-value problems proceeds as in the two preceding examples.

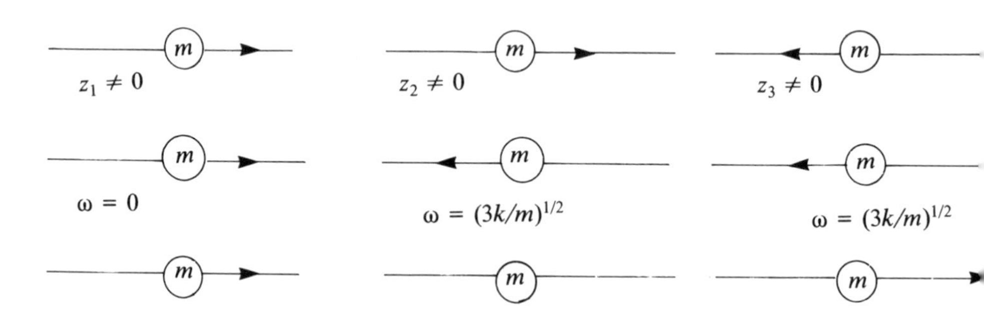

Figure 11.5. Normal modes of motion for the Lagrangian 11.42.

NOTES

Discussions and examples of normal coordinates may be found in the following references:

1. French, A.P. *Vibrations and Waves*. New York: Norton, 1971.
2. Pain, H.J. *The Physics of Vibrations and Waves.* New York: Wiley, 1968.
3. Morse, P.M. *Vibrations and Sound.* 2d ed. New York: McGraw Hill, 1948.
4. Symon, K.R. *Mechanics*. 3d ed. Reading MA: Addison Wesley, 1971.
5. Applications of the theory may be found in works on molecular vibrations and on vibrations of crystal lattices, for example, in treatments of infrared and Raman spectra and of phonon spectra.

PROBLEMS

11.1. Three masses slide without friction on three parallel rods that lie in the xy-plane. Two of the rods bear masses m; the rod between these two, equidistant from the two, bears a mass $2m$. The middle mass is connected to the others by two equal massless springs whose unstretched length is less than the separation of the rods. Calculate normal frequencies and normal coordinates and sketch them by means of arrows.

11.2. The system described in Problem 11.1 begins as three equal masses at rest on their rods. At $t = 0$ the middle mass is hit by an equal mass sliding at velocity v_0 along the middle rod; this mass sticks to the target, making a mass $2m$. How does the system move thereafter?

11.3. Calculate normal frequencies and coordinates for a system that is like the third example in the text (Lagrangian of Equation 11.42), except that there are four rods at the corners of a square and four equal masses, each connected by springs to its two neighbors.

11.4. Calculate normal coordinates and frequencies for two equal masses bearing equal and opposite electric charges and sliding, subject only to Coulomb force, on two straight rods that pass each other at angle θ and distance D.

11.5. Solve the same problem for two equal, oppositely charged, masses sliding without friction on two parallel planes. Then calculate their motion after they have been started opposite each other with initial velocities at right angles to each other but equal in magnitude.

11.6. Mass m_1 hangs from the ceiling by a massless spring of spring constant k and unstretched length S_0. Mass m_2 hangs from m_1 by a similar spring. Assuming vertical motion of both masses, calculate normal frequencies and normal coordinates.

11.7. Three rods parallel to the x-axis intersect the yz-plane at the vertices of an equilateral triangle; there is a mass m sliding without friction on each. A fourth rod, parallel to the three, intersects the yz-plane at the center of the triangle formed by the three; it bears a mass $3m$ that slides without friction and is connected to the three masses by three equal springs whose unstretched length is less than the separation of the rods. Calculate normal frequencies and coordinates, and sketch the normal modes of motion.

11.8. The system of Problem 11.7 consists initially of the three masses m coupled to a central mass $2m$, all at rest. At $t = 0$ a mass m moving along the central rod with velocity v hits the central mass and sticks to it. How does the system move thereafter?

12

Finite One-Dimensional Periodic Systems, Difference Equations

In this chapter we shall discuss certain systems, mechanical and electrical, whose properties are periodic or repetitive along the length of the systems. We shall begin with systems of finite length and shall later go on to discuss infinite systems. We shall find that, in defining normal coordinates for periodic systems, we are describing wavelike motions; the study of waves in such systems will give new insights into the properties of the systems.

Our first example is a system of N equal masses constrained to move only in the y-direction, with equilibrium positions along the x-axis at separation D, and with neighboring masses connected by springs of unstretched length $<D$ and effective spring constant k. Thus there are masses at all the positions $x = nD$, $n = 1 \ldots N$, with adjacent masses connected by springs that are all alike, and with the end masses, 1 and N, connected by similar springs to the points (0,0) and ($[N+1]D$,0). We shall denote by y_n the y-coordinate of the nth mass, and shall assume that $y_n \ll D$, so the vibrations of the chain of masses will be small vibrations about equilibrium. Although there are no masses at $x = 0$ and x

$= (N+1)D$, we shall take $y_0 = y_{N+1} = 0$ because the end springs are attached to fixed points at $y = 0$.

The kinetic energy of this system comes from the motion of the masses in the y-direction; they are constrained from moving in the x- or z-direction. Thus

$$T = \tfrac{1}{2}m \sum_{n=0}^{N+1} \dot{y}_n^2 = \tfrac{1}{2}m \, \tilde{\dot{y}} \, \dot{y}, \tag{12.1}$$

where the sum can include the terms $n = 0$ and $n = N + 1$, or not, since these y are constrained to be zero.

The potential energy comes from the stretching of the springs connecting adjacent masses. All these springs have the effective spring constant k. Thus,

$$V = \tfrac{1}{2}k \sum_{n=1}^{N+1} (y_n - y_{n-1})^2 = \tfrac{1}{2} \tilde{y} \, V y, \tag{12.2}$$

where

$$V = k \begin{pmatrix} 2 & -1 & 0 & 0 & . & . & 0 \\ -1 & 2 & -1 & 0 & . & . & 0 \\ 0 & -1 & 2 & -1 & . & . & 0 \\ . & 0 & -1 & 2 & -1 & . & . \\ . & . & . & . & . & 2 & -1 \\ . & . & . & . & . & -1 & 2 \end{pmatrix}. \tag{12.3}$$

Here, as in the case of T, we can include rows and columns for $n = 0$ and $n = N+1$, or we can omit them. Let us omit them in both cases; the matrices will all have N rows and/or N columns and, in particular, the V matrix is $N \times N$. Now, having factored the spring constant k out of the matrix V, let us factor it also out of the eigenvalues of V; let

$$K = kw, \tag{12.4}$$

and let us solve for the possible values of w. The secular equation can be written

$$D_N = 0, \tag{12.5}$$

where D_N is the $N \times N$ determinant

$$D_N = \begin{vmatrix} 2-w & -1 & 0 & 0 & 0 & 0 & 0 \\ -1 & 2-w & -1 & 0 & . & . & . \\ 0 & -1 & 2-w & -1 & . & . & . \\ . & . & . & . & 2-w & -1 & . \\ . & . & . & . & -1 & 2-w & -1 \\ 0 & 0 & 0 & 0 & 0 & -1 & 2-w \end{vmatrix}. \tag{12.6}$$

Our task now is to evaluate the determinant D_N. Using minors of the first column, we find that

$$D_N = (2\text{-}w)\, D_{N-1} + D'_{N-1}, \tag{12.7}$$

where D_{N-1} is just like D_N, but of size $(N-1)\times(N-1)$, and D'_{N-1} is the $(N-1)\times(N-1)$ determinant

$$D'_{N-1} = \begin{vmatrix} -1 & 0 & 0 & 0 \,.. & 0 & 0 & 0 \\ -1 & (2-w) & -1 & 0 \,.. & . & . & . \\ 0 & -1 & (2-w) & -1 \,.. & . & . & . \\ . & . & . & . \,.. & (2-w) & -1 & 0 \\ 0 & . & . & 0 \,.. & -1 & (2-w) & -1 \\ 0 & . & . & 0 \,.. & 0 & -1 & (2-w) \end{vmatrix} = -D_{N-2} \tag{12.8}$$

The last step comes from expanding D'_{N-1} by minors of its first row. Thus Equation 12.7 becomes

$$D_N - (2-w)D_{N-1} + D_{N-2} = 0. \tag{12.7$'$}$$

This is an example of a sort of equation known as a *difference equation;* such equations are closely analogous to differential equations. In Equation 12.7$'$ the independent variable is N and the unknown function of N is D_N. The equation is called a "difference" equation because it involves finite differences of the unknown, at discrete values of N. A so-called first difference of D_N is $D_N - D_{N-1}$; a second difference is $(D_N - D_{N-1}) - (D_{N-1} - D_{N-2}) = D_N - 2D_{N-1} + D_{N-2}$, a difference of first differences. Equation 12.7′ is a *second-order* difference equation because, to write it in terms of differences of D_N, one needs to use second differences; the points at which D enters the equation cover a range of 2, from N-2 to N, with three different values of D appearing.

Like a second-order differential equation, a second-order difference equation has a general solution that contains two arbitrary constants and that therefore requires two independent additional facts for its unique determination. By inspecting Equation 12.7′ one can see that knowing the value of D at two neighboring values of N, for instance, determines D at all other integer values of N, for the equation will directly determine D at the next higher and next lower points, and then a new choice of N in the equation will permit the next step upward or downward, etc. We shall use the values of D_N at $N = 1$ and $N = 2$ (a 1×1 and a 2×2 determinant) to determine the solution of Equation 12.7′. First, however, we have to find the general solution.

Equation 12.7′ is an example, also, of a *linear* difference equation, because the unknown function D_N appears only to the zeroth and first powers. In fact, because it appears only to the first power, the equation is also *homogeneous*. Furthermore, it has constant coefficients; N appears only in D_N, not in the coefficients. Thus Equation 12.7′ is a difference equation of the simplest type, linear, homogeneous, with constant coefficients, the analogue of linear homogeneous differential equations with constant coefficients. In the latter case we normally assume an exponential form for the solution; here we try a power-law form

$$D_N = E^N, \tag{12.9}$$

where we shall seek to find the constant E that satisfies the equation. Substitution of E^N into Equation 12.7′ gives

$$E^2 - (2\text{-}w)E + 1 = 0, \tag{12.10}$$

which has the two roots

$$E = 1 - \tfrac{1}{2}w + [w(\tfrac{1}{4}w - 1)]^{1/2},$$
$$1/E = 1 - \tfrac{1}{2}w - [w(\tfrac{1}{4}w - 1)]^{1/2}. \qquad (12.11)$$

These roots are known to be reciprocals because the last term of Equation 12.10 equals 1. So the Nth power of either E or $1/E$ will satisfy Equation 12.7′. The structure of the equation is such that any linear combination of such terms, with constant coefficients, also satisfies the equation:

$$D_N = aE^N + bE^{-N}. \qquad (12.12)$$

We shall take this as the general solution, for it has the correct number of constants.

These constants can be evaluated by use of two known values of D. As already stated, we shall use D_1 and D_2 for initial conditions; they are the 1×1 and the 2×2 determinants

$$D_1 = aE + b/E = 2 - w,$$
$$D_2 = aE^2 + b/E^2 = (2 - w)^2 - 1. \qquad (12.13)$$

In solving for a and b it is convenient to use Equation 12.10 to express w in terms of E rather than the contrary:

$$w = 2 - E - 1/E. \qquad (12.10')$$

Then, in terms of E,

$$a = E/(E - 1/E) = E^2/(E^2 - 1);$$
$$b = (1/E)/(1/E - E) = -1/(E^2 - 1). \qquad (12.14)$$

Thus,

$$D_N = (E^{N+2} - E^{-N})/(E^2 - 1) = E^{-N}(E^{2N+2} - 1)/(E^2 - 1). \qquad (12.15)$$

The last form is a convenient one to use in the secular Equation 12.5. The roots of $D_N = 0$ are those of $E^{2N+2} - 1 = 0$, unless the denominator $E^2 - 1$ may have roots that cancel some of these; E^{-N} cannot be zero, as one can see from 12.10. Thus E has to be one of the $2N+2$ roots of 1, but not ± 1, which are the roots of the denominator. Thus we can write

$$E_n = e^{\pi i n/(N+1)}, \; n = 1, 2, \; \ldots, N, N+2, N+3, \; \ldots, \; 2N+1. \tag{12.16}$$

These values of E can now be substituted into Equation 12.10′ to give the possible values of w and hence the eigenvalues of the matrix V:

$$\begin{aligned} K_n/k = w_n &= 2 - e^{\pi i n/(N+1)} - e^{-\pi i n/(N+1)} \\ &= 2 - 2\cos\pi n/(N+1) = 4\sin^2\pi n/2(N+1). \end{aligned} \tag{12.17}$$

Thus the normal frequencies $\omega_n = (K_n/m)^{1/2}$ are given by

$$\omega_n = 2(k/m)^{1/2}\sin\pi n/2(N+1). \tag{12.18}$$

In determination of w, K, and ω, only half of the n-values of Equation 12.16 are needed, for the values above $N+1$ duplicate the effect of those below that value; thus, in Equations 12.17 and 12.18, $n = 1, 2, 3, \ldots, N$. The $N \times N$ matrix V must have N eigenvalues, as we know; we now have N of them, all different, so we can be confident that we have found them all.

Now, knowing the values of w, we can substitute them into the linear equations for the components of the eigenvector R. $(V/k)R = wR$ gives:

$$\begin{aligned} (2\text{-}w)R_1 - R_2 &= 0 \\ -R_1 + (2\text{-}w)R_2 - R_3 &= 0 \\ -R_2 + (2\text{-}w)R_3 - R_4 &= 0 \\ -R_{N-2} + (2\text{-}w)R_{N-1} - R_N &= 0 \\ -R_{N-1} + (2\text{-}w)R_N &= 0. \end{aligned} \tag{12.19}$$

All these equations except the first and the last have the form

$$R_j - (2\text{-}w)R_{j-1} + R_{j-2} = 0, \qquad (12.20)$$

and by defining $R_0 - R_{N+1} = 0$, we can subsume the first and last equations also under the form of Equation 12.20. This is simply the same difference equation, for R_j, that we have already solved for D_N. Its general solution is given by Equation 12.12, *mutatis mutandis*, but now the boundary conditions are different:

$$R_0 = R_{N+1} = 0. \qquad (12.21)$$

From Equations 12.12 and 12.16, then, we can write for the nth column of R

$$R_{jn} = AE_n^{-j} + BE_n^{-j} = Ae^{\pi inj/(N+1)} + Be^{-\pi inj/(N+1).} \qquad (12.22)$$

Then $R_{j0} = 0$ implies

$$A = -B,$$

or

$$R_{jn} = 2iA \sin\pi nj/(N+1) = R_{nj}. \qquad (12.23)$$

This solution already satisfies $R_{j,N+1} = 0$, so that requirement provides no new information. Indeed, the eigenvalue problem determines eigenvectors, at most, up to a factor, so we can learn the value of A only by normallizing the nth column of R. While normalizing a column of R, we can also check the orthogonality of different columns as a partial check on our work. So we write

$$\sum_{n=1}^{N} R_{jn}R_{kn} = \sum_{n=0}^{N} R_{jn}R_{kn}$$

$$= -4A_jA_k \sum_{n=0}^{N} \sin\pi nj/(N+1) \sin\pi nk/(N+1), \qquad (12.24)$$

where we have allowed for the possibility that the constant A may be different in different columns of R, and have included (zero) terms for $n = 0$ in the sum over n. The reason for doing so is that the sum is most readily evaluated by the sine functions' again being expressed as exponentials. The exponentials that are present when terms are multiplied out are all roots of 1, and when $n = 0$ terms are included, they sum to zero because the roots of 1 do so (compare Equation 5.23, et seq.). The only exponentials that do not sum to zero are those that have zero exponents; these sum to $N+1$. The result of these arguments, as can be verified explicitly, is that the sum 12.24 vanishes when $j \neq k$, and equals $-2A_j^2(N+1)$ when $k=j$. Setting this latter sum equal to 1, then, we have

$$A_j = -i/[2(N+1)]^{1/2}, \qquad (12.25)$$

and

$$R_{jn} = R_{nj} = [2/(N+1)]^{1/2} \sin\pi nj/(N+1). \qquad (12.26)$$

The matrix R has now been made orthogonal; incidentally, it is symmetric, also. So

$$y = Rz \text{ and } z = Ry. \qquad (12.27)$$

Thus

$$\begin{aligned} y_n &= [2/(N+1)]^{1/2} \sum_j z_j \sin\pi jn/(N+1); \\ z_j &= [2/(N+1)]^{1/2} \sum_n y_n \sin\pi nj/(N+1). \end{aligned} \qquad (12.28)$$

Each z_j oscillates with its own frequency $\omega_j = 2(k/m)^{1/2} \sin\pi j/2(N+1)$, so the general time-dependent y_n (displacement of the nth mass) is the sum above, with z_j in each term being written as a sinusoidal function of the appropriate frequency, with two arbitrary constants. In order to study the normal modes of motion, we assume that only one z_j is nonzero, and ask how the various y_n behave in such a case. For example, the first normal mode has frequency $\omega_1 = 2(k/m)^{1/2} \sin\pi/2(N+1)$, the lowest of the ω_j, and in that mode

$$y_n = [2/(N+1)]^{1/2} c_1 \sin\pi n/(N+1) \cos(\omega_1 t + \phi_1). \qquad (12.29)$$

The motion of the y_n is shown schematically in Figure 12.1.

The second mode has frequency ω_2, which is almost twice ω_1; in it the y_n are

$$y_n = [2/(N+1)]^{1/2}\, c_2 \sin 2\pi n/(N+1) \cos(\omega_2 t + \phi_2), \tag{12.30}$$

depicted in Figure 12.2

The highest (Nth) mode has a frequency almost $2(k/m)^{1/2}$; the y_n are

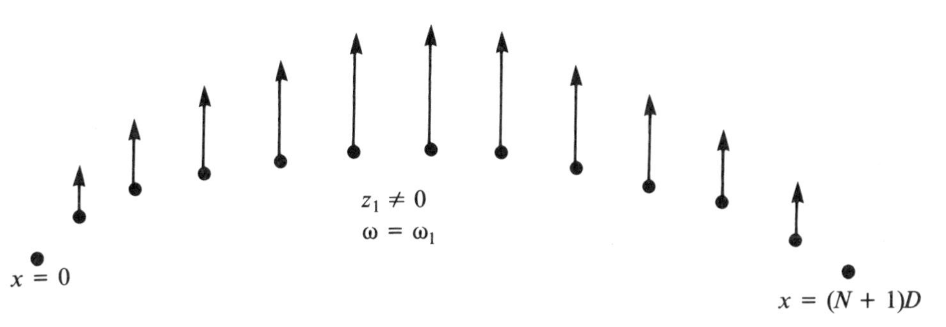

Figure 12.1. Lowest normal mode of the chain of masses.

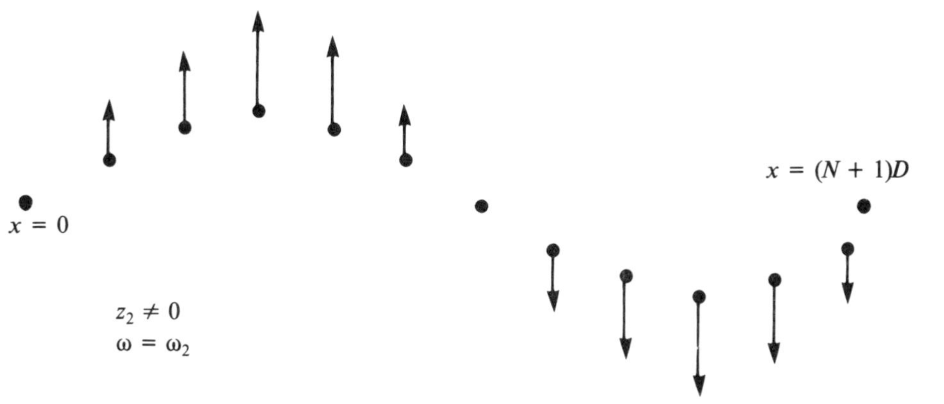

Figure 12.2. Second normal mode of the chain of masses.

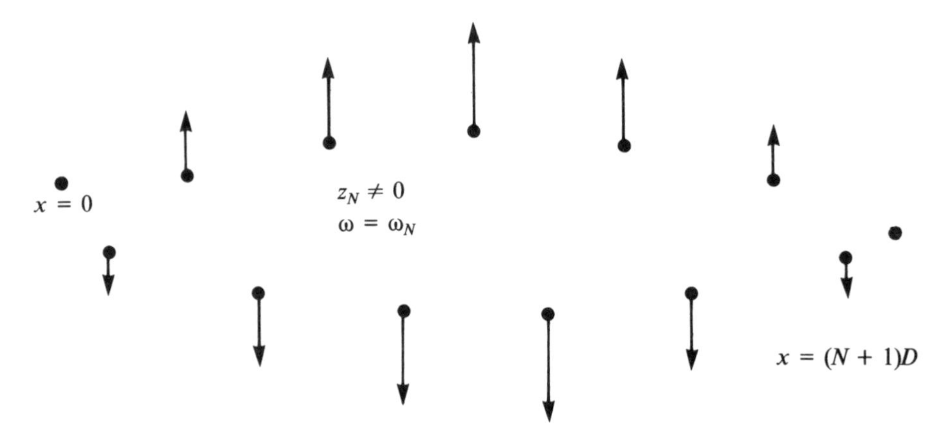

Figure 12.3. Highest normal mode of the chain of masses.

$$y_n = [2/(N+1)]^{1/2} \sin \pi n N/(N+1) c_N \cos(\omega_N t + \phi_N), \tag{12.31}$$

and are depicted in Figure 12.3.

So we see that each normal mode of motion is a standing wave with nodes at $x = 0$ and $x = (N+1)D$. Thus the jth standing wave, with j half-waves between these points, has wavelength

$$\lambda_j = 2(N+1)D/j. \tag{12.32}$$

The frequencies of these waves are related in a more complicated way, as shown in Equation 12.18. If N is large, the first few frequencies are proportional to the sines of small numbers, so they approximately form an arithmetic progression; ω_j is almost proportional to j when $j \ll N$. But as j becomes comparable to N the rate of increase of successive frequencies with increasing j decreases; at the maximum value of $j, j = N$, ω_j expressed as a function of j has almost leveled off.

The phase velocity of one of the waves is

$$v_j = \omega_j \lambda_j/2\pi = D(k/m)^{1/2} \frac{\sin \pi j/2(N+1)}{\pi j/2(N+1)}. \tag{12.33}$$

This phase velocity is nearly constant at $D(k/m)^{1/2}$ for the longest waves, $j \ll N$. As j becomes comparable to N, though, the velocity decreases, reaching a minimum value slightly higher than $(2/\pi)D(k/m)^{1/2}$. Figure 12.4 shows ω_j and v_j plotted against j.

Let us now suppose that all the masses are subjected to oscillating forces, of a common frequency ω; the force acting on the nth mass will be expressed as

$$\begin{aligned} F_n(t) &= A_n \cos(\omega t + \phi_n) \\ &= Re(A_n e^{i\omega t}). \end{aligned} \tag{12.34}$$

Equation 9.51 shows what linear combinations of the F_n act on the individual normal coordinates z_j. In the present case, however, the matrices M and S are both equal to the unit matrix I, and the masses have been kept with the accelerations. Thus,

$$\begin{aligned} m\ddot{z}_j + K_j z_j &= f_j(t) = \sum_n R_{jn} F_n(t) \\ &= [2/(N+1)]^{1/2} \sum_n F_n(t) \sin \pi nj/(N+1) \\ &= [2/(N+1)]^{1/2} e^{i\omega t} \sum_n A_n \sin \pi nj/(N+1), \end{aligned} \tag{12.35}$$

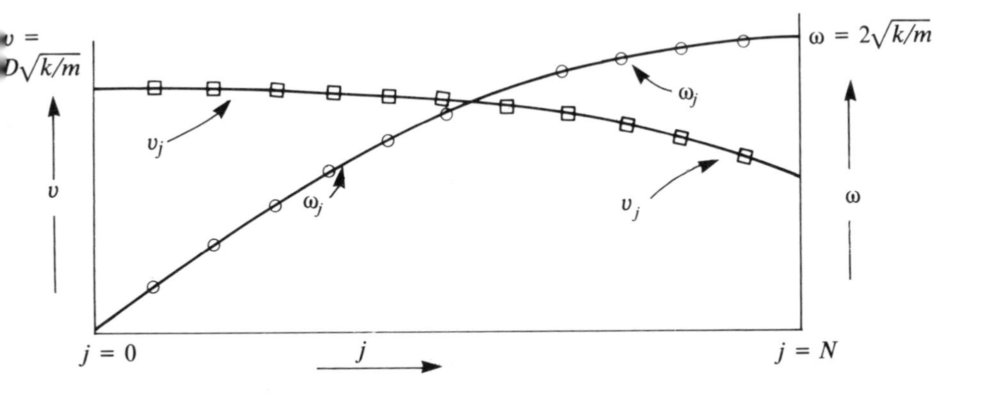

Figure 12.4. Frequencies and phase velocities of the first to N-th normal modes of the chain of masses.

where in the last step we have written the complex forces (as we have done in earlier chapters) without indicating real parts.

If the friction in the system really is zero, there will be no damped-out transient part of the response when forces like those in Equation 12.35 are turned on; the solutions of both the homogeneous and the inhomogeneous equations will have undying amplitudes. We shall assume, however, that there is a little friction, after all—too little to affect the steady-state response noticeably, but enough to have eliminated the transient. As indicated in Chapter 9, also, some friction is needed in order to move poles into the upper half-plane. Then we can take the steady-state response to sinusoidal forces of frequency ω to be

$$z_j(t) = [2/(N+1)]^{1/2}\, e^{i\omega t} \frac{\sum_n A_n \sin \pi nj/(N+1)}{m(\omega_j^2 - \omega^2)};$$

$$\dot{z}_j(t) = [2/(N+1)]^{1/2}\, i\omega e^{i\omega t} \frac{\sum_n A_n \sin \pi nj/(N+1)}{m(\omega_j^2 - \omega^2)}. \tag{12.36}$$

With the aid of Equation 12.27, these imply

$$y_r(t) = \sum_j R_{rj} z_j(t) = (e^{i\omega t}/m) \sum_{jn} R_{rj} R_{jn} A_n/(\omega_j^2 - \omega^2),$$

$$\begin{aligned} \dot{y}_r(t) &= \sum_j R_{rj} \dot{z}_j(t) = (i\omega e^{i\omega t/m}) \sum_{jn} R_{rj} R_{jn} A_n/(\omega_j^2 - \omega^2) \\ &= e^{i\omega t}\, [2/(N+1)] \sum_{nj} A_n\, [\sin \pi nj/(N+1)] \\ &\times [\sin \pi jr/(N+1)]/(i\omega m - iK_j/\omega). \end{aligned} \tag{12.37}$$

These responses can be most easily interpreted if one assumes that only one A_n is nonzero, i.e., that the only applied force acts on one of the masses, the nth, and that the entire motion of the system is a response to that one oscillating force. Then the sum over n reduces to one term in Equations 12.36 and 12.37, and the relative excitation of the various normal modes depends on two circumstances, the relative closeness of ω to the various ω_j (which determines which denominators in Equation 12.36 are small) and the relative sizes of the various sine functions with the n value of the mass that is subjected to the force and the j values of the various normal modes. If we can choose which mass to act on with a force of given amplitude A and frequency ω, we can produce maximum amplitude of a given standing wave by shaking a mass near an antinode of that wave (where the sine function is large) and using a given frequency, we can most strongly excite those waves with frequencies near the applied frequency. Although Equations 12.36 and 12.37 seem to predict infinite amplitude when ω equals one of the ω_j, there should be resistive terms in the denominators which will limit the magnitude of the response.

If one shakes one mass, n, with frequency ω and wish to investigate the relative amplitudes with which the various masses vibrate in response, one uses Equations 12.37 with the sum over n limited to one term. Even with this limitation the relationships are complicated; the excitations of different normal modes depend on the factors already mentioned, and then in addition the different normal modes involve the various y_k to different extents.

The impedance of the normal mode j at the point n is defined as the ratio of the complex force applied (only) at point n to the complex velocity of the jth normal coordinate. Thus

$$Z(z_j, n) = [\tfrac{1}{2}(N + 1)]^{1/2}\, i(\omega m - K_j/\omega)/[\sin \pi nj/(N + 1)]. \tag{12.38}$$

A given normal mode has its lowest impedance at its antinodes, where the sine function equals one; its impedance approaches infinity at the nodes. There being no dissipation in the system, the impedances are always purely reactive (imaginary); any energy put into the system is returned to the source. Each normal mode, at any given point of excitation (except a node) resembles a series LC circuit; its impedance is zero at its natural frequency. This means that, if it really could be free of dissipation, it could move at that frequency without any applied force.

The impedance of the system at a given point n, including the contributions of all the normal modes, is the ratio of A_n to $\dot{y}_n$, when no other forces are acting except the one that acts on the nth mass:

$$Z(n) = \frac{2/(N+1)}{\sum_j [\sin^2 \pi nj/(N+1)]/(i\omega m - iK_j/\omega)}, \qquad (12.39)$$

again purely reactive.

Having now discussed the chain of masses in terms of normal modes of motion, let us look again at the Lagrangian and the equations of motion expressed in terms of the original coordinates y_n:

$$L = \tfrac{1}{2}m \sum_n \dot{y}_n^2 - \tfrac{1}{2}k\, \Sigma_n (y_n - y_{n-1})^2; \qquad (12.40)$$

$$m\ddot{y}_n + k\,[(y_n - y_{n-1}) - (y_{n+1} - y_n)] = F_n(t); \qquad (12.41)$$

$$m\ddot{y}_n - k(y_{n+1} - 2y_n + y_{n-1}) = F_n(t).$$

Let us now specialize by setting $F_n(t) = 0$, and assuming a solution with the time dependence $y_n(t) = Y_n e^{i\omega t}$. Then $\ddot{y}_n = -\omega^2 y_n$, and the equation of motion reduces to

$$k[y_{n+1} - (2-\omega^2/\omega_0^2)y_n + y_{n-1}] = 0, \qquad (12.42)$$

where $\omega_0 = (k/m)^{1/2}$. This is the same difference equation satisfied by D_N(Equation 11.7′) and R_j (Equation 12.20), with w replaced by ω^2/ω_0^2. But w was defined as K/k, the eigenvalue divided by the spring constant; this is equal to $(K/m)/(k/m)$, or ω^2/ω_0^2, where ω is one of the eigenfrequencies. The boundary conditions on Equation 12.42 are $y_0 = y_{N+1} = 0$, which are the same as those on R_j in Equation 12.20. When solving for R, we found that satisfying the boundary condition at 0 led to the other one's being satisfied automatically. The reason for this was that we already knew the eigenvalues of V, and by using them in the solution for R we got columns of R that had integer numbers of half-wavelengths in the length of the string. In our present work we can assume that we have no idea of the possible frequencies ω with which the system can move in SHM, and can then find them by fitting both the boundary conditions that apply to the y_n. Thus we could have worked directly with the original coordinates y_n and arrived at the same normal modes of motion that we derived by a process of diagonalizing a matrix. This procedure might have been simpler in this case than the one that we used first, but the calculation of normal coordinates and normal frequencies is often very enlightening.

Figure 12.5 shows a finite periodic electrical system. In fact, we shall show that it is the electrical equivalent of the chain of masses just discussed. In Chapter 1 we arrived at the following equivalences:

$$\begin{gathered} x \text{ is equivalent to } Q; \\ \dot{x} \text{ is equivalent to } I; \\ m \text{ is equivalent to } L; \\ k \text{ is equivalent to } 1/C; \\ \text{Force is equivalent to emf}; \\ T \text{ is equivalent to } W_L = \tfrac{1}{2}LI^2; \\ V \text{ is equivalent to } W_C = \tfrac{1}{2}Q^2/C. \end{gathered} \tag{12.43}$$

This list suggests that, in the circuit depicted, the Lagrangian is

$$J = W_L - W_C = \tfrac{1}{2}L \sum_n \dot{Q}_n^{\,2} - (1/2C) \sum_n (Q_n - Q_{n-1})^2, \tag{12.44}$$

and that therefore

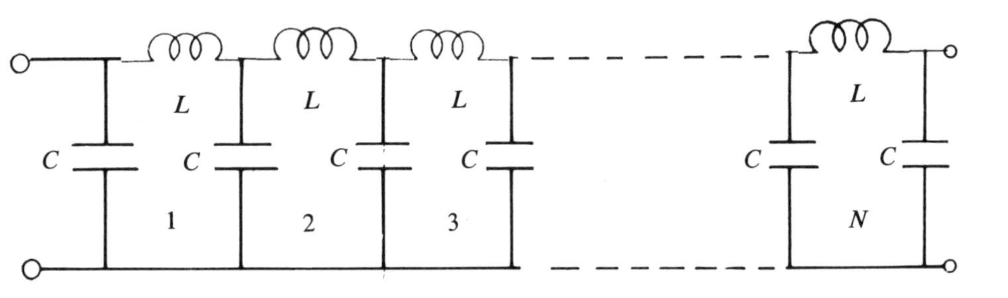

Figure 12.5. Finite periodic ladder network.

$$\frac{d}{dt}\frac{\partial J}{\partial \dot{Q}_n} - \frac{\partial J}{\partial Q_n}$$
$$= 0 = L\ddot{Q}_n + (1/C)\,[(Q_n - Q_{n-1}) - (Q_{n+1} - Q_n)], \tag{12.45}$$

the correctness of which is substantiated by direct application of Kirchhoff's laws to the loop currents and their time integrals Q_n.

These are the same equations, mutatis mutandis, that apply to the chain of N masses. Furthermore, if we let loops 0 and $N+1$ be the open loops at the ends of the chain of loops, we have $Q_0 = Q_{N+1} = 0$, like the corresponding ys. Thus, as claimed, the LC ladder network shown in Figure 12.5 is the electrical analog of the chain of masses, being its equivalent in all respects. So the frequency spectrum of the two is the same, as is their response to applied forces—which, in the circuit, would be produced by generators F_n in series with the inductances in the various loops.

This result suggests looking at other electrical networks that may not have mechanical analogs. The one in Figure 12.6 is such a network. Again taking $Q_0 = Q_{N+1} = 0$, we can write the Lagrangian

$$J = \tfrac{1}{2}L \sum_{n=1}^{N+1} (\dot{Q}_n - \dot{Q}_{n-1})^2 - \tfrac{1}{2} \sum_{n=1}^{N} Q_n^2/C, \tag{12.46}$$

and the equations of motion

$$L[(\ddot{Q}_n - \ddot{Q}_{n-1}) - (\ddot{Q}_{n+1} - \ddot{Q}_n)] + Q_n/C = 0,$$

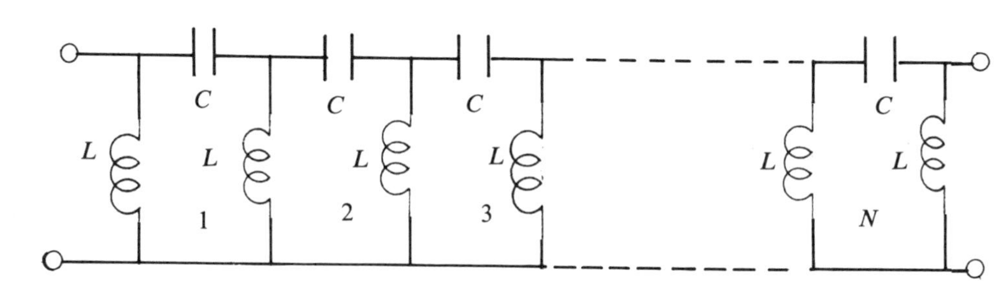

Figure 12.6. Network of Figure 12.5 with L and C interchanged.

or

$$Q_{n+1} - (2 - \omega_0^2/\omega^2)Q_n + Q_{n-1} = 0, \tag{12.47}$$

where the last form results from the assumption of sinusoidal time-dependence of some unknown frequency ω, and $\omega_0^2 = 1/LC$. This is the old familiar difference equation, again subject to $Q_0 = Q_{N+1} = 0$. But now the term ω^2/ω_0^2 has been replaced by ω_0^2/ω^2, so whatever values of the first ratio were needed to satisfy Equation 12.42 or its electrical analog, these same values of the inverted ratio apply here. Thus,

$$\omega_j = \tfrac{1}{2}\omega_0[\sin \pi j/2(N + 1)], \tag{12.48}$$

and each normal mode of motion is the same exactly as the corresponding one was before (as it must be, from comparison of Equation 12.47 and 12.42), except that its frequency is now given by Equation 12.48 instead of Equation 12.18. Thus, anomalously, the longest wavelength λ_1 corresponds to the highest frequency $\omega_1 = \omega_0/2\ [\sin\pi/2(N + 1)]$, much higher than ω_0, and the shortest wavelength, with one half-wave per section of the network, corresponds to the lowest frequency, slightly above $\omega_0/2$. So the phase velocities, expressed in sections per second, behave in a very anomalous way, and whatever the type and size of the hardware used in the construction of the network, the phase velocities of the longest waves are likely to be greater than the speed of light, c. Such a result seems to be violating the relativistic limitation on the speed of signals, but in fact it is not. The propagation of waves on such a network is extremely dispersive; i.e., the phase velocity is very strongly dependent on frequency—in contrast to the first case studied in this chapter, in which the phase velocity (shown in Figure 12.4), changes only slightly from the lowest to the highest frequency. Such strong dispersion means that a pulse, started at one end of the network as a signal being sent to the other end, gets drastically distorted en route because its various pure-frequency components travel at vastly different speeds and thus combine to give very different sums at different points along the way. Therefore a calculation of how fast the very front edge of the pulse travels (which is the signal velocity) is difficult and does not have to lead to a result greater than c.

Aside from the complication of dispersion, the circuit theory used here has the limitations discussed near the end of Chapter 7.

NOTES

1. A famous book on waves in periodic systems is Brillouin, L. *Wave Propagation in Periodic Structures.* New York: McGraw Hill, 1946; Dover Reprint, 1953.
2. Similar systems are discussed in books on lattice vibrations, for example, Seitz, F. *Modern Theory of Solids*. New York: McGraw Hill, 1940.

PROBLEMS

12.1 Generalize from Problem 11.3 to the case of N rods at the vertices of a regular N-gon, with each mass connected to its two neighbors (i.e., the chain of masses detached from the walls and closed on itself). Check the normal frequencies and normal modes of motion against the special cases $N = 2, 3, 4$.

12.2 The chain of masses discussed in the text is set in motion by a unit impulse applied to the nth mass at $t = 0$. How does the system move thereafter?

12.3 This chapter has treated the transverse vibrations (in the y-direction only) of the chain of masses. How do the longitudinal vibrations resemble these, and how are they different? Compare normal frequencies, normal modes of motion, and phase velocities.

12.4 A chain of $2N-1$ masses constrained to move only in the y-direction, is cut just after the Nth mass; all the masses and springs for $x > (N+1)\ D$ are replaced by two chains of masses, so the original chain is now branched in the middle. Assuming that each branch has the same length and the same number of masses that the original half-chain had, determine what must be the masses and spring constants in the new branches in order that the branched chain shall have all the normal frequencies and modes of motion that the original chain had. Because the new system contains $3N - 1$ masses, it must have also N modes that the old system lacked. Describe these new modes qualitatively.

12.5 A uniform horizontal tube of cross-sectional area A is connected with a row of equally spaced vertical tubes of the same cross-sectional area. The system is filled to height h above its bottom with a nonviscous, incompressible liquid that behaves as does the liquid in Problem 1.11. Analyze the small vibrations of this system about its equilibrium configuration in which the liquid reaches the same level in all the tubes. Assume that the horizontal tube is closed at both ends.

13

Infinite One-Dimensional Periodic Systems—Characteristic Impedance

This chapter is a continuation of the last one, in which the periodic systems are infinitely long—strictly speaking, semi-infinite, having a beginning but no end. We can get an indication of what to expect by letting our chain of masses, or the equivalent ladder network with series L and shunt C, become infinitely long by addition of more and more sections like those already there. Thus we let $N \rightarrow \infty$, without m or k (or L or C) changing. Then the spectrum of normal frequencies more and more densely fills the interval from 0 to $2\omega_0$, while ω_0 remains constant at $(k/m)^{1/2}$ or $(\mathrm{LC})^{-1/2}$. It seems reasonable to suppose that a semi-infinite system can oscillate at any frequency in this range, that the discrete spectrum becomes continuous in the limit. We shall see that this expectation is correct. Similarly, the series-C–shunt-L network, when infinitely long, permits any frequency above $\omega_0/2$, as might be expected from Equation 11.48.

Because they are more varied, we shall consider mainly electrical rather than mechanical networks, with the periodic ladder structure. Figure 13.1 shows a general specimen, as it functions at some particular frequency. The impedances Z_1 and Z_2 might be

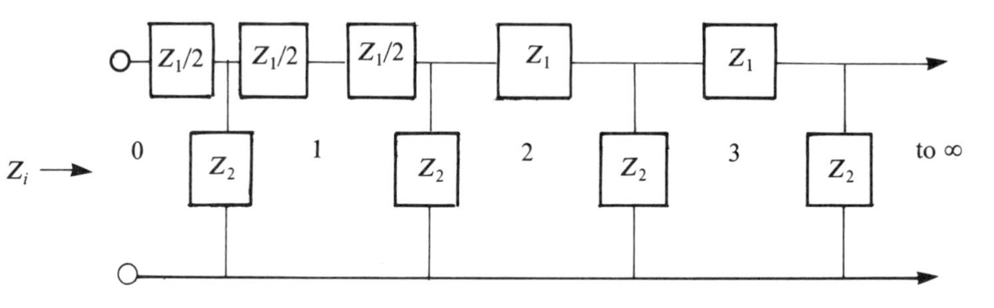

Figure 13.1. Semi-infinite ladder network with input impedance Z_i.

any complex functions of frequency. The first closed loop has the series impedance Z_1 drawn as a series combination of two impedances $Z_1/2$. The reason we depict it so is that we can thus more easily see how to calculate the input impedance Z_i. From the diagram we can see that Z_i consists of $Z_1/2$ in series with the parallel combination of Z_2 and ($Z_1/2$ in series with Z_i):

$$Z_i = Z_1/2 + \frac{Z_2(Z_i + Z_1/2)}{Z_2 + Z_i + Z_1/2}. \tag{13.1}$$

Here we have explicitly used the assumed semi-infinite length of the network to write the impedance of the network with the first T-section removed equal to the same Z_i that is the input impedance of the entire network including the first T-section; the network is not changed by removal of any finite number of sections. This fact has given us Equation 13.1, which implies

$$\begin{aligned} Z_i^2 &= Z_i Z_2 + \tfrac{1}{4} Z_1^2; \\ Z_i &= \pm[Z_1(Z_2 + \tfrac{1}{4} Z_1)]^{1/2}. \end{aligned} \tag{13.2}$$

The correct sign of the square root has to be chosen by some other criterion. For instance, if Z_i has a real (resistive) part, that part must be positive. A negative resistance in this situation would mean that the network is continuously giving up energy to whatever generator is connected to its input—an incredible state of affairs. When Z_i is pure imaginary we shall have to look further to discover its correct sign.

If either Z_1 or Z_2 contains resistance, Z_i necessarily does, too, for any resistances in the circuit will surely dissipate some energy. The more interesting cases are the idealized ones in which Z_1 and Z_2 are both purely reactive. Then no energy can be dissipated in the network, but Z_i can *still* contain resistance. The explanation of this fact lies in the possibility that energy from the generator at the input may move on down the network to infinity; energy may be put in and never return, not because it is dissipated, but because it gets propagated along the network.

Let us illustrate this possibility by means of our familiar example, $Z_1 = i\omega L$ and $Z_2 = 1/i\omega C$. Then

$$Z_i = \pm[L/C - \tfrac{1}{4}\omega^2 L^2]^{1/2}, \tag{13.3}$$

which is real when $\omega < 2\omega_0 \equiv 2(LC)^{-1/2}$, and imaginary when $\omega > 2\omega_0$. Without even knowing the sign of the reactance at high frequencies, we can be sure that no voltage can be propagated when $\omega > 2\omega_0$, because at those high frequencies, the network returns all the energy fed into it. On the other hand, below $2\omega_0$, Z_i is pure resistance (and positive), which must mean that the network propagates these low frequencies without reflection.

Our other example from Chapter 12, with series C and shunt L, has $Z_1 = 1/i\omega C$ and $Z_2 = i\omega L$, so

$$Z_i = [L/C - 1/4\omega^2 C^2]^{1/2}, \tag{13.4}$$

which is real when $\omega > \frac{1}{2}\omega_0$ and imaginary when $\omega < \frac{1}{2}\omega_0$, where ω_0 is still defined as $(LC)^{-1/2}$. Thus, by an argument like the preceding one, all frequencies above $\frac{1}{2}\omega_0$ are propagated, whereas lower frequencies are reflected back to the input. These results bear out the expectations at the beginning of this chapter. In general, we can identify the so-called *pass band* of a periodic ladder network (i.e., the range of frequencies that can propagate down the network) with those frequencies that make Z_i real. When the network contains no resistance, the Z_i^2 of Equation 13.2 is real, and Z_i is then either real or imaginary: there is either total propagation or total reflection. When Z_1 or Z_2 contains resistance, in general Z_i has both real and imaginary parts. There is always some reflection and some dissipation, and it is harder to discover what propagation occurs. But at all frequencies there is attenuation of the signal as it travels down the network because some energy is dissipated in every section.

The input impedance of a semi-infinite ladder network is sometimes called its *characteristic impedance.* A ladder network of finite length is sometimes used in a situation in which pulses must travel from one end to the other without being reflected upon reaching the far end. Then one tries to terminate the network in its characteristic impedance, for in that case all the currents and voltages in the network will be just as if the network had infinite length. The terminating impedance is the same as that of an infinite network like the finite one that is really there, so any signal that gets that far will simply propagate into the terminating impedance without reflection. The practical difficulty with this attempt is that the characteristic impedance of a network, given by Equation 13.2, has a frequency dependence that is hard to duplicate with standard hardware, so any realizable termination will match the network impedance at only one or two frequencies. Other frequencies will be reflected more or less strongly—and, of course, frequencies outside of the pass band will be reflected totally. Regardless of these difficulties, termination of a network by its characteristic impedance is an example of impedance-matching as this subject was referred to in Chapters 3 and 7. Elimination of reflection is the way to get maximum power transfer, for it amounts to preventing reflection of energy that arrives at the terminating impedance and, thus, dissipating all the energy in that impedance.

In order to analyze more fully ladder networks containing only reactances, let us define loop currents and apply Kirchhoff's laws, assuming that the only emf in the system is an input voltage $Fe^{i\omega t}$. Then, letting the voltage source be in loop 0, we have for $n = 1, 2, \ldots,$

$$-Z_2 I_{n-1} + (Z_1 + 2Z_2) I_n - Z_2 I_{n+1} = 0,$$

or (13.5)

$$I_{n-1} - (2+S)\, I_n + I_{n+1} = 0,$$

where $S = Z_1/Z_2$, and is real. Also we have at the input

$$F = I_0 Z_i,$$

and (13.6)

$$F = I_0(Z_2 + Z_1/2) - I_1 Z_2.$$

The difference Equation 13.5 has the same form that we encountered in Chapter 12, with S replacing $-w$. Thus the general solution is

$$I_n = AE^n + BE^{-n}, \tag{13.7}$$

where E and $1/E$ are the two roots of

$$E^2 - (2+S)E + 1 = 0 = (1/E)^2 - (2+S)(1/E) + 1.$$

That is, (13.8)

$$E = 1 + \tfrac{1}{2}S \pm \sqrt{S(1+\tfrac{1}{4}S)};$$

$$1/E = 1 + \tfrac{1}{2}S \mp \sqrt{S(1+\tfrac{1}{4}S)}.$$

The form of Equation 13.7 should be valid for all I_n beginning with I_0. To fix the constants A and B we use Equation 13.6:

$$I_0 = A + B = F/Z_i;$$

(13.9)

$$I_1 = AE + B/E = I_0(Z_2 + Z_1/2 - Z_i)/Z_2 =$$

$$I_0(1+\tfrac{1}{2}S - Z_i/Z_2).$$

Using Equation 13.2, we can write

$$Z_i/Z_2 = \pm \sqrt{S(1+\tfrac{1}{4}S)}. \tag{13.10}$$

Now we can resolve the ambiguity of sign in Equation 13.8. Let us define E in such a way that $I_1 = I_0 E$, that is

$$E = 1 + \tfrac{1}{2}S - Z_i/Z_2. \qquad (13.11)$$

This choice then leads to

$$B=0;\ A=F/Z_i = I_0;$$
$$I_n = I_0E^n = FE^n/Z_i. \qquad (13.12)$$

It remains now to interpret this solution in the various cases that may occur in a network comprising only reactances. The condition for the pass band (i.e., for frequencies that will be transmitted) is that input impedance Z_i is real (and, incidentally, positive). This condition implies that Z_i/Z_2 is imaginary and positive if Z_2 is a negative reactance, negative if Z_2 is positive imaginary. Thus the pass band corresponds to S lying in the range from -4 to 0, as implied by Equation 13.10

$$\text{Pass band: } -4 < S < 0. \qquad (13.13)$$

Thus, the network will not transmit if $S < -4$ or $S > 0$; these are the "stop bands".

$$\text{Stop band: } S < -4, \text{ or } S > 0. \qquad (13.14)$$

Let us look first at the case $Z_1/Z_2 = S < -4$. Now Z_i/Z_2 is real, and $1 + \frac{1}{2}S$ is negative. E and $1/E$ are both real and negative and, in view of its function in Equation 13.12, E must be the root that has absolute value less than 1, so that,

$$-1 < E < 0. \qquad (13.15)$$

Otherwise I_n would have larger and larger peak values as n increased, which cannot happen. In this case, then, in Equation 13.11, Z_i/Z_2 has to be negative, so Z_i is a reactance with sign opposite that of Z_2. This condition resolves the ambiguity of sign in Equation 13.2 in this case, $S < -4$.

If, instead, signals are attenuated because $S > 0$ (Z_1 and Z_2 are reactances of the same sign), $1 + \frac{1}{2}S$ is positive and Z_i/Z_2 is again real. Then the root that is less than 1 is the one to be identified with E, or Z_i/Z_2 must be real and positive. Thus both the cases in which signals are reflected back to the input instead of being propagated have the input impedance Z_i being a reactance *with the same sign as* Z_1, whether that is or is not the sign of Z_2. And I_n is proportional to the nth power of a number between -1 and 1.

In the pass band, Z_1/Z_2, or S, is between -4 and 0, Z_i is real and positive, and so Z_i/Z_2 is imaginary, having the sign of Z_1. Thus E and $1/E$ have equal real parts and opposite imaginary parts:

$$1/E = E^*,$$

whence (13.16)

$$| 1/E | = | E | = 1.$$

Let

$$E = e^{i\theta}, \tag{13.17}$$

so

$$I_n = I_0 \, e^{in\theta}, \tag{13.18}$$

where

$$\cos\theta = 1 + \tfrac{1}{2}S;$$

$$\sin\theta = -Z_i/iZ_2 = \pm[-S(1 + \tfrac{1}{4}S)]^{1/2}. \tag{13.19}$$

Thus $\sin\theta$, hence θ itself, has a sign opposite that of X_1, where $Z_1 = iX_1$—because Z_1 and Z_2 have opposite signs. Putting the time dependence back in now, we find that

$$I_n(t) = I_0 \, e^{i(n\theta + \omega t)}. \tag{13.20}$$

The wavelength of this wave, in segments, is $2\pi/|\theta|$; if each segment occupies D cm, the wavelength in cm is

$$\lambda = 2\pi D/|\theta|. \tag{13.21}$$

The phase velocity, in cm/sec, then should be

$$v_\phi = \omega\lambda/2\pi = D\omega/|\theta|. \tag{13.22}$$

Note that the content of Equations 13.17–13.20 is unchanged by replacement of θ by $\theta + 2k$, where k is any integer. Thus θ, hence v_ϕ, is ambiguous. One normally takes $|\theta|$ to be as small as possible, whether θ is positive or negative. However, the particular situation that we have been studying, in which a generator at one end of the network produces the signal that is propagated down the network, would seem to be one in which the phase velocity is $D\omega/|\theta_-|$, where θ_- is the negative value of θ with the smallest absolute value—negative because the wave is propagating toward positive (increasing) n, so $n\theta$ has to decrease when n increases, so as to compensate for an increase of ωt. Thus, in this situation, v_ϕ should be

$$\begin{aligned} v_\phi &= -D\omega/\theta_- \text{ for } \theta<0\ (X_1 < 0); \\ v_\phi &= D\omega/(2\pi - \theta) \text{ for } \theta > 0\ (X_1 > 0). \end{aligned} \tag{13.23}$$

Not only are phase velocities ambiguous in this system because θ is ambiguous; they do not have much physical importance anyway, for they do not directly tell how fast signals travel, or how fast energy is transmitted, or how fast a group or packet of waves travels. This last rate of travel is known as the *group velocity,* v_g. The question depends on the behavior of combination of sinusoidal waves that are all traveling with different phase velocities: How fast is a point of constructive interference propagated?

Let us consider a disturbance $u(x,t)$, a function of time, t, and a space variable x (which might be a distance measured in cm, or might be a number of segments of a periodic structure, like our n). Let us take the Fourier transform of u as a function of x at $t = 0$ to be a function $a(k)$; here k is a real variable, *not* to be confused with a spring constant!

$$u(x,0) = (2\pi)^{-1/2} \int_{-\infty}^{\infty} a(k)e^{-ikx}\, dk. \tag{13.24}$$

We assume that $a(k)$ has a maximum at k_0, and that each function e^{-ikx} propagates like $e^{i[\omega(k)t-kx]}$. That is, each wavelength $\lambda = 2\pi/k$ has its own frequency $\omega(k)$ and propagates with a phase velocity $\omega(k)/k$, which is how much x has to increase, when t increases by 1, in order to hold the phase $\omega t - kx$ constant. Thus we are assuming that all the exponential components are traveling in the same direction, toward $+x$, with various phase velocities. The fact that ω/k is a function of k (or of ω) means that the propagation is dispersive, that the pulse or wave packet changes shape while it travels. Let us assume that $\omega(k)$ is a smooth-enough function of k to be well approximated in the region of k where $a(k)$ is large, by two terms of a Taylor expansions about k_0. Then, as a function of x and t, the disturbance u is

$$\begin{aligned} u(x,t) &= (2\pi)^{-1/2} \int a(k)\, e^{i[\omega(k)t-kx]}\, dk \\ &\simeq (2\pi)^{-1/2} \int a(k)\, e^{i(\omega_0 t-k_0 x)+i(k-k_0)[\omega'(k_0)t-x]}\, dk \\ &= (2\pi)^{-1/2}\, e^{i(\omega_0 t-k_0 x)} \int a(k)\, e^{i(k-k_0)[\omega'(k_0)t-x]}\, dk, \end{aligned} \tag{13.25}$$

where ω_0 means $\omega(k_0)$, and $\omega'(k_0)$ means $d\omega/dk$ at $k=k_0$.

The function of x and t in front of the integral has constant absolute value, so the integral itself contains the variation of $|u|$ with x and t. In that integral the quantities x and t appear only in the combination $\omega'(k_0)t - x$, in which the coefficient of t is independent of k. Thus, to the extent that the approximation is accurate, the wave group moves in such a way that unit increase of t, if accompanied by an increase of x equal to $\omega'(k_0)$, leaves the absolute value of u constant. The group moves with the group velocity

$$v_g = d\omega/dk \text{ at } k = k_0. \tag{13.26}$$

In our ladder network we can let x be Dn, the distance of the nth loop from the origin, then our k is $-\theta/D$ and the group velocity is

$$v_g = -d\omega/d(\theta/D) = -D\,\frac{d\cos\theta/d\theta}{d\cos\theta/d\omega} = \frac{2D\sin\theta}{dS/d\omega}. \tag{13.27}$$

Here we have used Equation 13.19. Now we recall that the sign of $\sin\theta$ is opposite that of X_1 (where $Z_1 = iX_1$), $S = Z_1/Z_2$, and $Z_2 = iX_2$, where X_2 and X_1 have opposite signs in the pass band. So, taking $S = X_1/X_2$, we have

$$dS/d\omega = (X_2X_1' - X_1X_2')/X_2^2 = (SX_1' - S^2X_2')/X_1, \quad (13.28)$$

where the prime means derivative with respect to ω. If these derivatives are positive (as, for example, in the cases of ωL and $-1/\omega C$), then $v_g > 0$—because $X_1 \sin\theta < 0$, $S < 0$, and $S^2 > 0$. So, regardless of the sign chosen for the ambiguous phase velocity, the group velocity of $e^{i(n\theta + \omega t)}$ is positive. If, in some other situation, the term in E^{-n} is present also, that part of the solution will have a negative group velocity in the pass band; it is possible to have wave-groups going in opposite directions at the same time.

The situation treated in this chapter, in which a voltage generator introduces a wave of frequency lying in the pass band, which wave travels down a semi-infinite network, is one that involves only traveling waves; any energy that is present travels only in one direction in the network. By contrast, the previous chapter dealt with standing waves, which can be understood as a manifestation of two traveling waves with the same amplitude and frequency, going in opposite directions:

$$\cos(\omega t - kx) - \cos(\omega t + kx) = 2 \sin\omega t \sin kx. \quad (13.29)$$

Thus the waves on the chain of masses or the electrical ladder network of finite length can be thought of as the result of a traveling wave getting reflected at the end, re-reflected at the opposite and, and so on, so that there are two such waves traveling in opposite directions and interfering in such a way as to produce zero amplitude at the nodes and double amplitude at the antinodes.

In case a sinusoidal wave travels down the network and gets *partially* reflected at the far end, with the reflected wave being absorbed at the starting point, the waves in the network will be partly standing and partly traveling:

$$\cos(\omega t - kx) - A\cos(\omega t + kx)$$
$$= 2A \sin\omega t \sin kx + (1 - A)\cos(\omega t - kx). \quad (13.30)$$

Here we assume $A < 1$; the partial reflection results from partial energy loss upon reflection, an effect that one would expect from a terminating impedance that contains some resistance. Similarly, if the terminating impedance is a pure reactance the reflection should be total.

We can treat these questions more precisely by reverting to the general solution in Equation 13.7 and using the fact that, in the pass band, $E = e^{i\theta}$. It is convenient to terminate the ladder network with an nth loop containing Z_2 in common with the $(n\text{-}1)$st loop, and $Z_1/2$ in series with a terminating impedance Z'. Then we can solve for A, B, and the current in the last loop, from the zeroth, $(n\text{-}1)$st, and nth loop equations. We get for the amplitude ratio of waves going in the two directions

$$\frac{A}{B} = e^{-2in\theta}\frac{i\sin\theta - P}{i\sin\theta + P} = e^{-2in\theta}\frac{Z_i + Z'}{Z_i - Z'} \tag{13.31}$$

where $P = Z'/Z_2$. Also, the input impedance is no longer given by Equation 13.2; it is

$$Z_i' = F/I_0 = F/(A + B) = Z_2 \sin\theta \, \frac{P\cos n\theta - \sin\theta \sin n\theta}{P\sin n\theta + \sin\theta\cos n\theta}$$

$$= Z_i \frac{Z_i \sin n\theta + iZ' \cos n\theta}{Z' \sin n\theta + iZ_i \cos n\theta} \tag{13.32}$$

Recall that a nondissipative network has both Z_1 and Z_2 pure imaginary, so S is real (and between -4 and 0). If the terminating impedance Z' is imaginary also, P is real, and the absolute value of A/B is unity. Reflection is total, and all that remains variable is the phase of the reflected wave relative to that of the incident wave, which phase determines the location of the nodes and antinodes of the standing wave. Note also that a real P renders the input impedance pure imaginary (because Z_2 is); the only way Z_i' can be real is for Z' to be the Z_i of Equation 13.2. If one holds the circuit parameters constant and varies n, the number of rungs in the ladder, the Z_i' of Equation 13.32 varies periodically, with the numerator and the denominator alternately going through zero. Since n is a discrete variable, probably neither member of the fraction will ever be precisely zero, but each may become quite small. This behavior is typical of the impedance of a system that supports a standing wave. (The previous chapter dealt with a similar case.)

In ending this chapter, we shall refer again to the last topic in Chapter 7, in which the output voltage and current from a four-terminal network were expressed in terms of the input voltage and current. This procedure, although more complicated than our usual method of solving for currents in terms of input voltage, has the advantage of being, in a passive circuit, independent of what the circuit is connected to. In Equation 7.25 we wrote output quantities, V_n and I_n, as linear combinations of input V_1, I_1, with coefficients A, B, a, b. We can now express the relationship in matrix notation:

$$\begin{pmatrix} V_n \\ I_n \end{pmatrix} = \begin{pmatrix} A & B \\ a & b \end{pmatrix} \begin{pmatrix} V_1 \\ I_1 \end{pmatrix} \text{ or } W' = UW, \tag{13.33}$$

where the definitions of the matrices W', U, and W are obvious. We showed in Chapter 7 that the determinant of the matrix U equals 1:

$$\det U = 1, \tag{13.34}$$

or U is said to be *unimodular*. In this respect it is like some of the rotation matrices, but they were real and were the reciprocals of their transposes, whereas U is in general complex (and, incidentally, frequency-dependent).

The particular advantage of this matrix representation of the effect of a four-terminal circuit is that circuits connected in cascade (output of one to input of the next) have an overall effect represented by a matrix which is the product of the matrices of the constituent circuits. We shall not pursue this matter here; it is discussed more fully in Brillouin's book.[1]

NOTES

1. Brillouin, L. *Wave Propagation in Periodic Structures*. New York: McGraw Hill, 1946; Dover Reprint, 1953.

2. See also: French, A.P. *Vibrations and Waves.* New York: Norton, 1971; and Zeines, B. *Introduction to Network Analysis* Englewood Cliffs, NJ: Prentice-Hall, 1967.

PROBLEMS

Calculate the pass bands of the following semi-infinite ladder networks. In each case determine how to make the pass band as wide as possible, and as narrow.

13.1

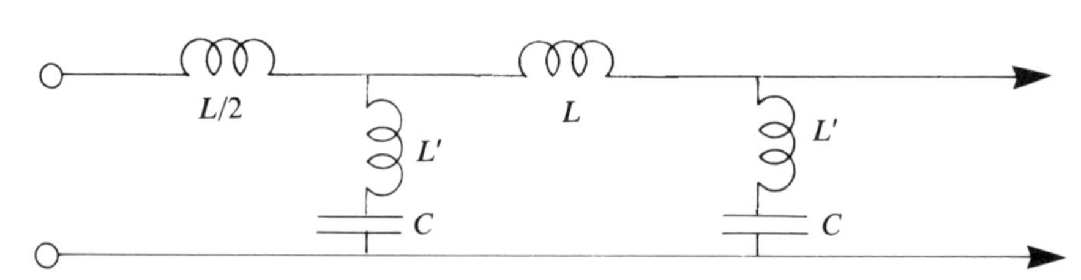

13.2

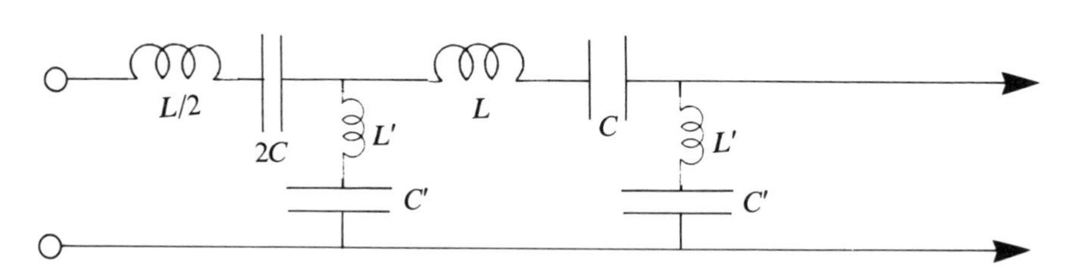

13.3

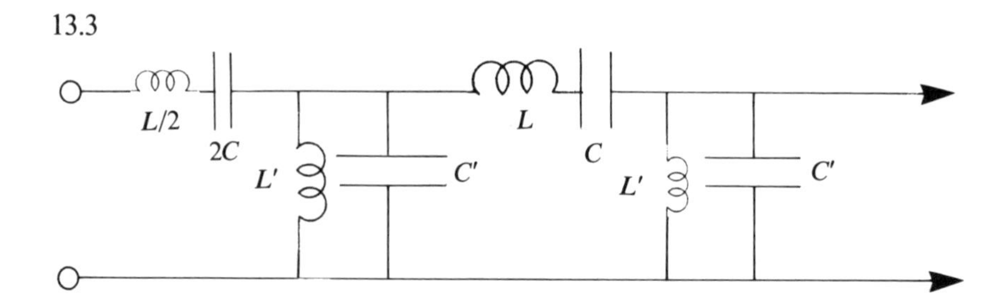

13.4 Let U be the (unimodular) matrix, first mentioned in Chapter 7, that gives output voltage and current in terms of input voltage and current for a certain four-terminal network that contains no resistance. Note that U is a function of frequency. Now imagine N of these networks connected in cascade, with the output of one constituting the input of the next, and the output of the Nth being the input of the first. What condition must U satisfy? Explain why, *carefully*. How might this condition be used in a search for the frequencies that can exist in this circuit that is closed on itself? Compare the roles of U and U^{-1} in this problem.

14

Continuous Systems, Wave Equation, Lagrangian Density, Hamilton's Principle

The presence of waves on periodic networks suggests that such networks may be related to continuous systems such as stretched strings on which also waves can be propagated. We explore this possibility by letting the segments of a periodic system become smaller and more numerous until, in the limit, the system is continuous. We concentrate our attention on the mechanical chain of masses discussed in Chapter 12, which becomes a continuous string in the limit, but shall later comment briefly on the limiting forms of other periodic systems.

The chain of masses has masses m spaced by distance D in the x-direction; the system has N masses and total length $(N + 1)D$ between its fixed ends. The spring constant of each spring is k_0 and its equilibrium length is S_0. Each spring is stretched to length D when the masses are at rest, thus having tension $k_0 (D-S_0) = Dk_0(1-S_0/D) = Dk$, where k is the effective spring constant for lateral displacements. So we can speak of a mass-per-unit-length,

$$\mu = m/D, \tag{14.1}$$

and a tension

$$\mathcal{T} = k_0(D - S_0) = kD. \tag{14.2}$$

We shall let m and D go to zero, and k and N go to infinity, in such a way that $\mu = m/D$, $\mathcal{T} = kD$, and $\Lambda = (N + 1)D$ are all constant. Let us do this in Equation 12.41, taking $F_n(t) = 0$. We divide the equation by D:

$$(m/D)\ddot{y}_n - kD\,\frac{(y_{n+1} - y_n)/D - (y_n - y_{n-1})/D}{D} = 0. \tag{14.3}$$

If we now express y as a function of x and t rather than n and t, we can write $\ddot{y}_n$ as $\partial^2 y/\partial t^2$; $y_{n+1} - y_n$ is now $y(x + D) - y(x)$, and $y_n - y_{n-1}$ is $y(x) - y(x-D)$. Thus, as we let the various parameters approach their limits, the complicated fraction approaches $\partial^2 y/\partial x^2$. Letting m/D be called μ, and kD, $\mathcal{T}$, we have

$$\mu\,\frac{\partial^2 y}{\partial t^2} = \mathcal{T}\,\frac{\partial^2 y}{\partial x^2},$$

or

$$\mu y_{tt} = \mathcal{T} y_{xx}, \tag{14.4}$$

or

$$v^2 y_{xx} = y_{tt}.$$

Here we have used the subscript notation for partial derivatives that was introduced in Chapter 5, and have abbreviated the combination of parameters in the last step:

$$v^2 = \mathcal{T}/\mu, \text{ or } v = \sqrt{\mathcal{T}/\mu}. \tag{14.5}$$

We shall call Equation 14.4 a *wave equation.* We already have reason to expect that it will describe wavelike disturbances, for its counterpart in the periodic system has wavelike solutions. In fact, we can verify this expectation immediately by working out the general solution of Equation 14.4. Let

$$u = x + vt,$$
$$w = x - vt, \tag{14.6}$$

and use these as new independent variables in place of x and t. Thus

$$\frac{\partial}{\partial x} = \frac{\partial \mu}{\partial x}\frac{\partial}{\partial u} + \frac{\partial w}{\partial x}\frac{\partial}{\partial w}; \quad \frac{\partial}{\partial t} = \frac{\partial u}{\partial t}\frac{\partial}{\partial u} + \frac{\partial w}{\partial t}\frac{\partial}{\partial w}; \text{ etc.} \tag{14.7}$$

Then Equation 14.4 becomes

$$\partial^2 y/\partial u \partial w = 0, \tag{14.8}$$

the general solution of which is

$$y = f(w) + g(u) = f(x - vt) + g(x + vt), \tag{14.9}$$

where f and g are any two functions that have first derivatives. Thus the wave Equation 14.4 is satisfied by any two disturbances traveling with speed v without changing size or shape, in opposite directions. We shall later study various properties of this general solution. First, however, we shall look at the limiting forms of potential and kinetic energy for the stretched string.

Kinetic energy is given in Equation 12.1. If we write D as Δx in the sum, we have

$$T = \tfrac{1}{2}\sum \mu y_t^2 \Delta x \rightarrow \tfrac{1}{2}\mu \int_0^{\Lambda} y_t^2 dx, \tag{14.10}$$

where the sum became an integral in the limit. Potential energy is given in Equation 12.2, and can be written as

$$v = \tfrac{1}{2}\mathscr{T}\sum_n \frac{(y_n - y_{n-1})^2}{D^2}\Delta x \rightarrow \tfrac{1}{2}\mathscr{T}\int_0^{\Lambda} y_x^2 dx \tag{14.11}$$

Thus we see that, for the continuous string, T and V are both expressed as integrals, over the length of the string, of quantities that might be called densities of kinetic and potential energies. The sum of these is then the total energy of the string:

$$E = T + V = \int_0^\Lambda \mathscr{E}\, dx = \int_0^\Lambda [\tfrac{1}{2}\mu y_t^2 + \tfrac{1}{2}\mathscr{T} y_x^2]\, dx, \quad (14.12)$$

where $\mathscr{E}$, the energy per unit length, is the integrand. Also, we should find that the string has a Lagrangian function, which ought still to be $T - V$:

$$L = T - V = \int_0^\Lambda \mathscr{L}\, dx = \int_0^\Lambda [\tfrac{1}{2}\mu y_t^2 - \tfrac{1}{2}\mathscr{T} y_x^2] dx, \quad (14.13)$$

where the Lagrangian density $\mathscr{L}$ is given by

$$\mathscr{L} = \tfrac{1}{2}\mu y_t^2 - \tfrac{1}{2}\mathscr{T} y_x^2. \quad (14.14)$$

It should be possible to derive equation(s) of motion from a Lagrangian, but the method for doing so is not obvious in the case of a continuous system. There would seem to be an uncountable infinity of coordinates, since to specify the configuration of the system at a given time one needs to give a function $y(x)$ which has an independent value at each point x. Although the dependence of L on the time derivatives of these coordinates seems reasonably clear, its dependence on the coordinates themselves is more complicated, and either case it is not obvious how one is to interpret the derivatives that enter Lagrange's equations.

This problem is most readily approached by means of Hamilton's principle, which was derived for a system of discrete particles in Chapter 9. As usually stated, this principle is

$$\delta \int_{t_1}^{t_2} L\, dt = 0 = \delta \int_{t_1}^{t_2}\int_0^\Lambda \mathscr{L}\, dx dt, \quad (14.15)$$

where the last form results from the fact that $\mathscr{L}$ itself is an integral over x. For greater generality, we shall deal with a three-dimensional system in which there is a dependent variable $U(x,y,z,t)$, and a Lagrangian density $\mathscr{L}$ which depends on U and all its first partial derivatives U_x, U_y, U_z, U_t. We shall let $u(x,y,z,t)$ be the particular function U that satisfies the laws of motion, subject to given conditions on the boundary of the space-time region in which we wish to know u; the given conditions will consist of specification of the function u itself on all points of the boundary.

Let us choose our space-time region to be a rectangular parallelepiped and a time interval $[t_1, t_2]$, and let $g(x,y,z,t)$ be any function that has its first derivatives and that vanishes on the boundary of the region. Then let the varied function U be

$$U(x,y,z,t) = u(x,y,z,t) + hg(x,y,z,t), \qquad (14.16)$$

where h is a real number. Then we shall interpret Hamilton's principle to say that

$$d/dh \int \mathscr{L}(U,U_x,U_y,U_z,U_t)\, dxdydzdt \Big|_{h=0} = 0. \qquad (14.17)$$

Since the limits do not involve h, we can differentiate under the integral sign, using the h-dependence of $\mathscr{L}$ as shown below:

$$\mathscr{L}(U,U_x,U_y, U_z,U_t)$$
$$= \mathscr{L}(u + hg, u_x + hg_x, u_y + hg_y, u_z + hg_z, u_t + hg_t). \qquad (14.18)$$

Thus Equation 14.17 gives

$$\int\left[\frac{\partial \mathscr{L}}{\partial U} g + \frac{\partial \mathscr{L}}{\partial U_x} g_x + \frac{\partial \mathscr{L}}{\partial U_y} g_y + \frac{\partial \mathscr{L}}{\partial U_z} g_z + \frac{\partial \mathscr{L}}{\partial U_t} g_t\right]_{h=0} \times dxdydzdt = 0$$

$$= \int\left[\frac{\partial \mathscr{L}}{\partial u} g + \frac{\partial \mathscr{L}}{du_x} g_x + \frac{\partial \mathscr{L}}{\partial u_y} g_y + \frac{\partial \mathscr{L}}{\partial u_z} g_z + \frac{\partial \mathscr{L}}{\partial u_t} g_t\right] dxdydzdt \qquad (14.19)$$

$$= \int\left[\frac{\partial}{\partial x}\left(g\, \frac{\partial \mathscr{L}}{\partial u_x}\right) + \frac{\partial}{\partial y}\left(g\, \frac{\partial \mathscr{L}}{\partial u_y}\right) + \frac{\partial}{\partial z}\left(g\, \frac{\partial \mathscr{L}}{\partial u_z}\right) + \frac{\partial}{\partial t}\left(g\, \frac{\partial \mathscr{L}}{\partial u_t}\right)\right.$$

$$+ g\left(\frac{\partial \mathscr{L}}{\partial u} - \frac{\partial}{\partial x}\frac{\partial \mathscr{L}}{\partial u_x} - \frac{\partial}{\partial y}\frac{\partial \mathscr{L}}{\partial u_y} - \frac{\partial}{\partial z}\frac{\partial \mathscr{L}}{\partial u_z} - \frac{\partial}{\partial t}\frac{\partial \mathscr{L}}{\partial u_t}\right)\Bigg]$$

$$\times\, dxdydzdt = 0$$

Here the first line has $d\mathscr{L}/dh$ written out, with the total dependence of $\mathscr{L}$ on h taken into account. The second line sets $h = 0$; it is understood that here and hereafter $\mathscr{L}$ is a function of u, u_x, u_y, u_z, u_t, that is, $h = 0$. The long integral in the third and fourth lines is simply a rewriting of the one above, preparatory to integration by parts. We now integrate the third line, doing first the variable of differentiation in each term. This first integration yields values of functions proportional to g, evaluated on the boundary. These contributions all vanish because g is defined to be zero on the boundary. Then we are left with the last line equal to zero:

$$0 = \int g\left(\frac{\partial \mathscr{L}}{\partial u} - \frac{\partial}{\partial x}\frac{\partial \mathscr{L}}{\partial u_x} - \frac{\partial}{\partial y}\frac{\partial \mathscr{L}}{\partial u_y} - \frac{\partial}{\partial z}\frac{\partial \mathscr{L}}{\partial u_z} - \frac{\partial}{\partial t}\frac{\partial \mathscr{L}}{\partial u_t}\right) \times dxdydzdt \qquad (14.20)$$

This integral has to be zero regardless of the function g, provided only that it is once-differentiable and vanishes on the boundary. The only way this can be true with such a great variety of possible functions g is that the integrand vanishes almost everywhere (which to a physicist means everywhere):

$$\frac{\partial \mathscr{L}}{\partial u} - \frac{\partial}{\partial x}\frac{\partial \mathscr{L}}{\partial u_x} - \frac{\partial}{\partial y}\frac{\partial \mathscr{L}}{\partial u_y} - \frac{\partial}{\partial z}\frac{\partial \mathscr{L}}{\partial u_z} - \frac{\partial}{\partial t}\frac{\partial \mathscr{L}}{\partial u_t} = 0. \qquad (14.21)$$

This is the form taken by Lagrange's equations for a single continuous system, or *field*, u, in three dimensions. It can be extended to a system of several fields, each varied by its own function $g(x,y,z,t)$, and in general interacting with each other (i.e., having coupled equations of motion arising from a function that contains products of different fields and of functions of these fields).

In our case the field u is y, the displacement of the string from equilibrium, and the independent variables are x,t instead of x,y,z,t. Furthermore, $\mathcal{L}$, given in Equation 14.14, depends only on derivatives of y and not on y itself. Applying Equation 14.21 to the $\mathcal{L}$ of Equation 14.14 we get

$$\frac{\partial}{\partial x}\frac{\partial \mathcal{L}}{\partial y_x} + \frac{\partial}{\partial t}\frac{\partial \mathcal{L}}{\partial y_t} = u y_{tt} - \mathcal{T} y_{xx} = 0, \tag{14.4'}$$

the same equation of motion as Equation 14.4.

This equation can be derived by more direct reasoning—though perhaps not quite so well justified—by consideration of the forces acting on a segment between x and $x + dx$. If one assumes that this segment has length dx (that is, it forms a small angle with the x-axis), then it has mass μdx, and the net force acting on it in the y-direction should equal $(\mu dx)\ddot{y}$:

$$(\mu\, dx)\ddot{y} = dF_y. \tag{14.22}$$

If no external force is acting on the string, dF_y acting on a segment dx is caused entirely by the neighboring parts of the string. At the left end, at x, the tension $\mathcal{T}$ is acting at an angle $\theta(x)$; at the right end, at $x + dx$, $\mathcal{T}$ is acting at an angle $\theta(x + dx)$. Thus the resultant force in the y-direction is $\mathcal{T}[\sin\theta(x + dx) - \sin\theta(x)]$, assuming that $\mathcal{T}$ is the same at both ends of the segment. If we assume further that θ is small enough for $\sin\theta$ to be replaced by $\tan\theta = y_x$, we can write

$$(\mu\, dx)\ddot{y} = (\mathcal{T} dx)\, \partial \tan\theta/\partial x,$$

or (14.4″)

$$\mu y_{tt} = \mathcal{T} y_{xx},$$

Equation 14.4 again.

We have an expression for energy per unit length, $\mathcal{E}$, in Equation 14.12. Let us now seek an expression for energy flow rate, i.e., the rate at which the string to the left of x does work on the string to the right of x. The rate at which a force **F** does work on an object on which it acts is $\mathbf{F}\cdot\mathbf{v}$, where $\mathbf{v}$ is the velocity of the object acted on. Here the string just above x has its velocity all in the y-direction; it is y_t. The y-component of force acting from below x is, as approximated above, $-\mathcal{T}y_x$. Thus the energy-flow rate, or power flow, is

$$W = -\mathcal{T}y_x y_t. \tag{14.23}$$

We can check this expression, and Equation 14.12 for $\mathcal{E}$, by verifying that energy is conserved. The rate of change of the amount of energy between $x = a$ and $x = b > a$ should be the rate at which energy is moving toward $+x$ at a, minus the rate at which it is moving toward $+x$ at b:

$$\frac{d}{dt}\int_a^b \mathcal{E}\, dx = W(a) - W(b) = -\int_a^b (\partial W/\partial x)dx. \tag{14.24}$$

Since the two integrals have to be equal regardless of their limits a and b, the integrands have to be equal:

$$\partial\mathcal{E}/\partial t = -\partial W/\partial x. \tag{14.25}$$

Using expressions 14.12 and 14.23 for $\mathcal{E}$ and W, we evaluate the derivatives indicated in Equation 14.25, getting

$$\mu y_t y_{tt} + \mathcal{T}y_x y_{xt} = \mathcal{T}(y_{xx}y_t + y_x y_{tx}). \tag{14.26}$$

According to Equation 14.4 the first terms on the two sides of Equation 14.26 are equal; the two second terms are identical. Thus Equation 14.26 is satisfied, and the expressions for $\mathcal{E}$ and W are consistent with energy conservation.

Equation 14.9 shows the general solution of the equation of motion 14.4. Let us evaluate $\mathcal{E}$ and W in terms of this general solution, $y = f(x - vt) + g(x + vt)$. Thus, using $v^2 = \mathcal{T}/u$,

$$\begin{aligned}\mathcal{E} &= \tfrac{1}{2}\mu v^2\,[g' - f']^2 + \tfrac{1}{2}\mathcal{T}[f' + g']^2 \\ &= \mathcal{T}\{[g'(x + vt)]^2 + [f'(x - vt)]^2\}.\end{aligned} \tag{14.27}$$

Also,

$$W = \mathcal{T}v\,[g' + f']\,[f' - g']$$
$$= v\mathcal{T}\{[f'(x - vt)]^2 - [g'(x + vt)]^2\}. \qquad (14.28)$$

Here the primes mean differentiation of each function with respect to the one variable on which it depends, that is, $f'(x - vt) = df(x - vt)/d(x - vt)$, and likewise for g'. It is noteworthy that neither $\mathcal{E}$ nor W contains an interference term that depends on both f and g. Each wave, the one, f, going toward $+x$, and the one, g, going toward $-x$, has its own energy density and power flow. Not surprisingly, the power flow of f equals v times the energy density of f, whereas for g the factor is $-v$:

$$W_+ = v\mathcal{E}_+$$
$$W_- = -v\mathcal{E}_-. \qquad (14.29)$$

Here W_+ and $\mathcal{E}_+$ mean those associated with the + directed wave, $f(x - vt)$, and the minus subscripts denote quantities associated with the − moving $g(x + vt)$. These relationships are as if the energy were a fluid of density $\mathcal{E}_\pm$, moving with velocity v or $-v$, so that its flow is what it would be for such a fluid, velocity times density.

Equation 14.4 contains second derivatives with respect to x and t. We can use arguments involving Taylor series in deciding what initial or boundary conditions are needed for unique specification of a solution. For example, knowing $y(x,0)$, y as a function of x at $t = 0$, is *not* enough, for if we attempt a Taylor series in t about $t = 0$, we lack the first-derivative term. Thus, we need $y_t(x,0)$ also. Having both y and y_t as functions of x at $t = 0$, we have the first two terms in a Taylor series in t, and $y(x,0)$ determines $y_{xx}(x,0)$, which determines $y_{tt}(x,0)$ via the wave equation. Then $y_t(x,0)$ determines y_{txx}, which, via the wave equation, gives y_{ttt}, etc. The whole Taylor series in t is determined from the two functions $y(x,0)$ and $y_t(x,0)$. Similar reasoning implies that $y(0,t)$ and $y_x(0,t)$ determine a solution. This procedure is not recommended as a way of getting solutions; it is used here merely to show what information is needed to determine a solution.

We shall now assume that we have complete initial information:

$$y(x,0) = q(x),$$
$$y_t(x,0) = p(x), \tag{14.30}$$

where p and q are known functions. We shall now embody this information in the general form $y = f(x - vt) + g(x + vt)$:

$$f(x) + g(x) = y(x,0) = q(x);$$
$$v[g'(x) - f'(x)] = y_t(x,0) = p(x). \tag{14.31}$$

Now letting $P(x)$ be the indefinite integral of $p(x)$, uncertain by an arbitrary constant C, we can integrate the second Equation 14.31 to get

$$g(x) - f(x) = [P(x) + C]/v. \tag{14.32}$$

Now, using Equation 14.31 again, and Equation 14.32, we find that

$$g(x) = \tfrac{1}{2}[q(x) + P(x)/v + C/v];$$
$$f(x) = \tfrac{1}{2}[q(x) - P(x)/v - C/v]. \tag{14.33}$$

These equations give the functional forms of f and g. However, we want the particular combination $f(x - vt) + g(x + vt)$:

$$y(x,t) = f(x-vt) + g(x+vt)$$
$$= \tfrac{1}{2}[q(x-vt) + q(x+vt) - P(x-vt)/v + P(x+vt)/v], \tag{14.34}$$

which determines y completely in terms of the initial conditions (Equation 14.30) involving the functions q and p.

In our discussion of the string, so far, we have not assumed that it was fixed at $x = 0$ and $x = \Lambda$. Even in our use of Hamilton's principle, where we assumed that y had certain specified values on the boundary of the region of interest in the x,t-plane, we did not assume that y was fixed at $x = 0$ and $x = \Lambda$. We merely assumed that, whatever functions (of t) y was on these boundaries, the variation of the integral left these functions unchanged. Thus all our results so far apply to either the finite or the infinite string.

The unconstrained infinite string can sustain any functions $f(x-vt)$ and $g(x+vt)$; the functions $q(x)$ and $p(x)$ that represent initial conditions can be any two continuous, differentiable functions. On the other hand, if both $y(t)$ and $y_x(t)$ are fixed at one point, the motion is fully determined; there is no freedom to choose initial conditions. The case in which y is fixed, but not y_x, is thus an intermediate case; it restricts but does not determine the motion. If $y(0,t) = 0$, we find that

$$g(vt) + f(-vt) = 0, \tag{14.35}$$

where vt can be any real number; thus, at any time,

$$g(x) + f(-x) = 0, \tag{14.36}$$

for all x. Thus, f and g are inverted mirror images of each other, reflected in the point $x=0$, at all times. Figure 14.1 shows an example of this situation. As the pulses in Figure 14.1 travel, the pulse labeled g will approach $x=0$; as it begins getting there, the negative pulse labeled f is arriving from the other side in such a way as to leave the string unmoved at $x=0$.

If, in addition, the string is fixed at $x=\Lambda$, we have

$$g(\Lambda + vt) + f(\Lambda - vt) = 0, \tag{14.37}$$

which implies that f and g at any given time have the same inverted-image relationship with respect to Λ as they have with

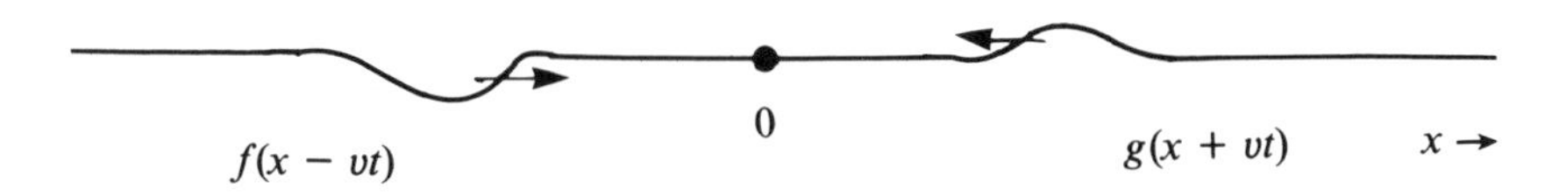

Figure 14.1. Pulses on a string that will leave the string stationary at x = 0.

respect to 0. These two restrictions still leave quite a lot of leeway in the choice of the function $y(x,t)$, despite the fact that fixing y at two values of x might appear equivalent to fixing y and y_x at one point—as is the case normally in the fixing of positions at two times instead of positions and velocities at one time (compare Hamilton's principle). In the present case, however, the mirror-image relationship with respect to two points puts a disturbance between $x=0$ and $x=\Lambda$ in the position of the observer between two parallel mirrors. Not only does that pulse have images reflected in the two fixed points 0 and Λ, but each image has an image, etc. That is, the image at $-x$ has an image in the point at Λ, at the position $2\Lambda+x$, twice inverted or uninverted, and so on. The result of these restrictions is that there is a periodic sequence of inverted pulses moving in the opposite direction and so timed as to cancel the first sequence at the two points 0 and Λ. Thus what appears between 0 and Λ is a periodic disturbance that leaves the string fixed at 0 and Λ and has a period equal to 2Λ (the spacing) divided by the speed v:

$$\tau = 2\pi/\omega_1 = 2\Lambda/v = 2\Lambda(\mu/\mathcal{T})^{1/2}. \qquad (14.38)$$

Although we have discussed this situation as if we had an infinite string that happened to be fixed at 0 and Λ, the usual case in which the string is so constrained is that of a finite string fixed at both ends, that is, the true limiting case of the finite chain of masses. On such a finite string the only part of the periodic sequences of pulses (described above) that actually exists at a given time is the part which at that time lies between 0 and Λ. The other periods of the periodic function of x are, as it were, ghosts which come into existence when they enter the interval $[0, \Lambda]$ and die when they leave it, the net effect of their life cycle being to produce in $[0, \Lambda]$ the same behavior that would occur on the infinite string fixed at those points. Each wave on the finite string gets reflected upside down from each end of the string, and every disturbance on the string is periodic in time.

We can get more insight into the properties of the finite string by referring to the normal coordinates and normal frequencies worked out in Chapter 12; these quantities are displayed in Equations 12.18 and 12.28 of that chapter. Using the relationships expressed in the first two equations of the present chapter and the following text, we can rewrite the expression for normal frequencies as follows:

$$\omega_n = 2\sqrt{\mathcal{T}/\mu}\,\frac{\sin\{\pi n/[2(N+1)]\}}{(\pi n/[2(N+1)]}\;\frac{\pi n}{2(N+1)D} \rightarrow \sqrt{\frac{\mathcal{T}}{\mu}}\,\frac{\pi n}{\Lambda}, \tag{14.39}$$

where the fraction of the form $\sin x/x$ goes to 1 because $N + 1$ becomes infinite while n remains finite, and, as before, Λ is the length of the string. Note that ω_1 in the limit agrees with what we called ω_1 in Equation 14.38, the frequency of the periodic disturbances on the finite string. In Equation 14.39 we are on the quasi-linear part of the sine function that gave the frequencies of the chain of masses. In our present limit the frequencies are equally spaced like those in a Fourier series, so any disturbance on the string *does* have to be periodic, with fundamental frequency ω_1. Now the frequencies have no upper limit; adjacent masses have acquired zero spacing, so there is no lower limit on wavelength. The string has an infinite number of mechanical coordinates, so there is an infinite number of normal frequencies and of normal coordinates.

The transformation matrix R was defined so that $y = Rz$. Now we know that the y_n have become a continuous set, and we expect that the z_j have become countably infinite in number. Thus the rows of R have become a continuum, and the columns have become infinite in number but not continuous. R_{nj} should now be written accordingly:

$$R_{nj} = \tilde{R}_{jn} \rightarrow R_j(x), \tag{14.40}$$

and

$$\begin{aligned} y(x, t) &= \sum_j R_j(x)\, z_j(t); \\ z_j(t) &= \int_0^\Lambda y(x,t)\, R_j(x)\, dx. \end{aligned} \tag{14.41}$$

The orthogonality of the matrix R now takes the form

$$\int_0^{\Lambda} R_j(x)\, R_k\, (x)dx = \delta_{jk};$$
$$\sum_j R_j(x)\, R_j(x') = \delta(x - x'). \tag{14.42}$$

Taking R_{nj} from Equation 12.26, we can write the orthogonality of different columns of R (Equation 10.41), as

$$\delta_{jk} = [2/(N+1)D] \sum_n \sin[\pi jnD/(N + 1)D]$$
$$\times \sin\, [\pi knD/(N + 1)D]\; \Delta(nD), \tag{14.43}$$

where $nD = x$, and $\Delta(nD) = \Delta x = D$, which cancels the D that has been put in the denominator to convert $N + 1$ to $(N + 1)D = \Lambda$. Thus the limiting form of Equation 14.43 is the (correct) equation

$$\frac{2}{\Lambda}\int_0^{\Lambda} \sin\frac{\pi jx}{\Lambda} \sin\frac{\pi kx}{\Lambda}\, dx = \delta_{jk}, \tag{14.44}$$

which implies that the limiting form of R_{nj} is

$$R_j(x) = \sqrt{\frac{2}{\Lambda}} \sin\frac{\pi jx}{\Lambda}, \tag{14.45}$$

and guarantees the correctness of the first Equation 14.42. The functions $R_j(x)$ are an orthonormal set in the interval $0 < x < \Lambda$, so if y is given by the first of Equations 14.41, the second has to give the coefficients z_j correctly (as stated, for instance, in Equation 14.7). Furthermore, substituting from the second Equation 14.41 for z_j in the first Equation 14.41 and reversing the order of sum and integral shows that the sum in the second Equation 14.42 is the identity operator, i.e., the delta function that it is supposed to be.

With the expression 14.45 for the elements of the transformation matrix R, we can write Equations 14.41 as

$$y(x,t) = \sqrt{\frac{2}{\Lambda}} \sum_j z_j(t) \sin(\pi jx/\Lambda);$$

$$z_j(t) = \sqrt{\frac{2}{\Lambda}} \int_0^\Lambda y(x,t) \sin(\pi jx/\Lambda)\, dx. \tag{14.41'}$$

Thus the normal coordinates z_j are coefficients in a Fourier-like series—Fourier-"like" because the interval $[0,\Lambda]$ is not an integer number of periods of the odd-numbered sine functions. The fundamental period is 2Λ, not Λ. However, the sine functions being used are orthogonal in the interval $[0,\Lambda]$, and they have the convenient property of vanishing at the ends of the string, so the $y(x,t)$ in Equation 14.41 can never violate the constraints $y(0,t) = y(\Lambda,t) = 0$.

We already know the general time dependence of the z_j; from Chapter 12, we know that z_j oscillates with frequency ω_j. Now ω_j is as given by Equation 14.39. The z_j thus are still expected to be normal coordinates of the system—a fact that we can verify by writing the Lagrangian with Equation 14.41′ incorporated:

$$\begin{aligned}\mathcal{L} &= \tfrac{1}{2} \int_0^\Lambda [\mu y_t^2 - \mathcal{T} y_x^2]\, dx \\ &= \frac{1}{\Lambda} \sum_{kj} \left\{ \mu \dot{z}_j \dot{z}_k \int_0^\Lambda \sin \frac{\pi jx}{\Lambda} \sin \frac{\pi kx}{\Lambda}\, dx \right. \\ &\quad \left. - \mathcal{T} z_j z_k \int_0^\Lambda \frac{\pi j}{\Lambda} \frac{\pi k}{\Lambda} \cos \frac{\pi jx}{\Lambda} \cos \frac{\pi kx}{\Lambda}\, dx \right\},\end{aligned} \tag{14.46}$$

where we have already written the integrals of double sums as double sums of integrals—assuming uniform convergence. We know that the integral of sines is $(\Lambda/2)\,\delta_{jk}$; without the extra constant factors in the integrands the integrals of cosines are the same. Thus

$$\mathcal{L} = \sum_j [\tfrac{1}{2}\mu \dot{z}_j^2 - \tfrac{1}{2}\mathcal{T}(\pi^2 j^2/\Lambda^2)\, z_j^2], \tag{14.47}$$

a sum of squares (uncoupled) which leads to the equations of motion:

$$\mu \ddot{z}_j + \mathcal{T}(\pi^2 j^2/\Lambda^2)\, z_j = 0, \tag{14.48}$$

whereby z_j moves in SHM with frequency

$$\omega_j = \frac{\pi j}{\Lambda}\sqrt{\frac{\mathcal{T}}{\mu}} = \frac{\pi j v}{\Lambda}, \tag{14.39$'$}$$

as in Equation 14.39. The coefficients z_j are indeed normal coordinates; they still have the significance that they had in Chapter 12 when they pertained to the discrete chain of masses. They are the time-dependent amplitudes of the various standing waves that can exist in the system.

One can now write the general motion of the finite continuous string, making explicit the dependence on both x and t:

$$y(x,t) = \sqrt{\frac{2}{\Lambda}}\sum_j [a_j \cos(\pi j v t/\Lambda) + b_j(\sin \pi j v t/\Lambda)] \sin(\pi j x/\Lambda), \tag{14.49}$$

where the a_j and b_j are the constants of integration, to be determined from initial conditions. Problem 14.3 involves expressing the standing waves in Equation 14.49 as sums of traveling waves of equal amplitude going in opposite directions. The possibility of so doing is a sign of the fact that a traveling wave is reflected at the end of the string with no loss of energy.

If, as previously, we assume that, in $0 < x < \Lambda$,

$$\begin{aligned} y(x,0) &= q(x), \\ y_t(x,0) &= p(x), \end{aligned} \tag{14.30$'$}$$

we can fit the solution 14.49 to these conditions. Thus,

$$\begin{aligned} y(x,0) = q(x) &= \sqrt{\frac{2}{\Lambda}} \sum_j a_j \sin \frac{\pi j x}{\Lambda}, \\ y_t(x,0) = p(x) &= \sqrt{\frac{2}{\Lambda}} \sum_j b_j \omega_j \sin \frac{\pi j x}{\Lambda} \end{aligned} \tag{14.50}$$

whence

$$a_j = \sqrt{\frac{2}{\Lambda}} \int_0^\Lambda q(x) \sin \frac{\pi j x}{\Lambda} dx;$$
$$b_j = \frac{1}{\omega_j} \sqrt{\frac{2}{\Lambda}} \int_0^\Lambda p(x) \sin \frac{\pi j x}{\Lambda} dx. \tag{14.51}$$

According to Equations 14.51, if the string is plucked, i.e., displaced and released from rest, only the a_j will be present, having the values that they have at $t = 0$. When the string is hit, starting from $y = 0$ with nonzero velocity, only the b_j will be present and the amplitudes of higher harmonies are likely to decrease rapidly with increasing j because of the denominators ω_j.

Note that, as in previous Green's function derivations, one can substitute the expressions for the a_j, b_j from Equation 14.51 into Equation 14.49, thus explicitly expressing $y(x,t)$ in terms of its initial conditions. On reversing the order of summation and integration, one gets:

$$y(x,t) = \int_0^\Lambda dx' [\mathcal{G}(x,x',t-t')\, y(x',t') + \mathcal{G}'(x,x',t-t')\, y_t(x',t')] \tag{14.52}$$

where

$$\mathcal{G}(x,x',t-t') = \sqrt{\frac{2}{\Lambda}} \sum_j \sin \frac{\pi j x}{\Lambda} \sin \frac{\pi j x'}{\Lambda} \cos \frac{\pi j v(t-t')}{\Lambda};$$
$$\mathcal{G}'(x,x',t-t') = \sqrt{\frac{2}{\Lambda}} \sum_j \frac{\Lambda}{\pi j v} \sin \frac{\pi j x}{\Lambda} \sin \frac{\pi j x'}{\Lambda} \sin \frac{\pi j v(t-t')}{\Lambda}. \tag{14.53}$$

Here we have generalized to the extent of imposing initial conditions at t' instead of 0. Note, also, that

$$\mathcal{G} = \frac{\partial \mathcal{G}'}{\partial (t-t')}. \tag{14.54}$$

The expressions in Equations 14.52 and 14.53 are not useful. Although one can formally sum $\mathcal{G}$ by expressing the trigonometric functions in terms of exponentials and summing the geometric series thus obtained, these series diverge and so do not lead to usable results. But the argument used here can in other cases lead to practical relationships—as will be shown in Chapter 16. Green's functions can be used not only to determine a system's response to an applied force (as we have done) but also to generate the development of a system's state from a given initial state.

Our present treatment of the string implies that there is no damping. Actual strings are often rather rapidly damped, as one realizes on hearing a piano or a guitar. In the next chapter we shall have a few comments on frictional forces.

We close this chapter with a brief discussion of musical scales. The subjective scale of perceived pitch is a logarithmic scale of frequency: two tones with a given frequency ratio are perceived to have a given interval of pitch, regardless of whether the frequency invloved are low, medium, or high. For example, a 2:1 ratio of frequency produces a perceived interval of an octave. Two tones an octave apart sound much alike, and, for purposes of harmony, are nearly equivalent. Other ratios produce other perceived intervals. In general, frequencies having ratios of small integers sound consonant when heard together. This fact is one of the bases on which is built the diatonic scale, which has for some centuries been pervasive in its influence in western music. (In a few cases, its influence has been felt through its deliberate avoidance by some composers.) This scale is defined by the frequency ratios between pairs of its tones. These ratios are shown in Figure 14.2 for the diatonic scale beginning on the tone C (middle C has a frequency in the range 261–264 Hz, and other Cs have frequencies of this value times powers of 2).

The scale is defined by the sequence of frequency ratios between adjacent tones, as shown in the column immediately to the right of the column of letter designations. As shown, these ratios imply others that often involve small integers. The impression of consonance and harmony results from simultaneous sounding of tones with frequency ratios involving integers up to about 6 (for example, 3/2, 5/4, 6/5), whereas ratios like 9/8 and 10/9 give an impression of mild dissonance, and 16/15 sounds quite disso-

C	D	E	F	G	A	B	C′	
1	9/8	5/4	4/3	3/2	5/3	15/8		C
	1	10/9	32/27	4/3	40/27	5/3	16/9	D
		1	16/15	6/5	4/3	3/2	8/5	E
			1	9/8	5/4	45/32	3/2	F
				1	10/9	5/4	4/3	G
					1	9/8	6/5	A
						1	16/15	B
							1	C′

Figure 14.2. Frequency ratios of tones of the diatonic scale. Each fraction is the ratio of the frequency of the tone at the head of the column to that of the tone at the right end of the row of the table.

nant. The smallest integers, such as 2/1 and 3/2, give combinations that are so harmonious that they sound rather "thin," but a combination like C,E,G,C, a so-called major chord, sounds harmonious and "full."

If one wished to use a diatonic scale starting on D instead of C, the scale would have the same intervals, hence the same frequency ratios, as in Figure 14.2. Clearly few of the tones in the C-scale could be used in the D-scale—only G and B. If one required a scale beginning on each tone of the C-scale and, then, still more scales based on the new tones that had to be introduced to provide the original seven, etc., each octave would have to contain hundreds of tones, many of which would be so close together in pitch that few listeners could tell them apart. As music became more complicated and composers wished to make many *modulations* (changes of key), it became necessary to arrive at a compromise that would approximately satisfy the demand for all possible diatonic scales without encumbering instruments with too many keys. The compromise that has been used almost universally since about 1700 is the so-called even-tempered scale. This is the familiar chromatic scale, which has twelve equal intervals in an octave; each frequency ratio of adjacent tones is $2^{1/12} \cong 1.059$. One needs very accurate perception to distinguish intervals based on this scale from those based precisely on a diatonic scale.

An important feature of the string clamped at both ends is its spectrum of equally spaced frequencies. If we call the lowest tone (one half-wave on the string) the *fundamental* and the higher frequencies *overtones,* the string is distinguished by all its overtones having frequencies that are integer multiples of its fundamental frequency. The effect of this spectrum is that the fundamental and the first five overtones are all in the diatonic scale based on the fundamental; if the fundamental is C, and primes denote higher octaves, the overtones are: first, C′; second, G′; third, C″; fourth, E″, fifth, G″; sixth, between A″ and B″. So the fundamental and the five lowest overtones are all part of the major chord based on the fundamental; only with the seventh frequency (sixth overtone) does a dissonant tone occur. As Equations 14.51 show, the relative amplitudes of various frequencies depend on how a string has been set in motion. In the case of a string driven by a continuing force (caused, perhaps, by a bow) these relative amplitudes depend also on the action of the driver. But, whatever the means of excitation, the sixth and higher overtones are likely to be very weak, so the sound of the string is unlikely to be dissonant. Other vibrating systems have different spectra of overtones; some of them, such as bells, are likely to sound somewhat dissonant whenever they are sounded.

NOTES

1. The author is indebted to the late M.J. Moravcsik for the method of displaying frequency ratios used in Figure 14.2.
2. Many books contain discussions of vibrating strings; for example, the books, already cited: Symon K.R. *Mechanics* 3d ed. Reading, MA: Addison Wesley, 1971; and Morse, P.M. *Vibrations and Sound* 2d ed. New York: McGraw Hill, 1948.
3. The application of Hamilton's principle to continuous systems and, in general, the mechanics of such systems, are discussed in Goldstein, and in the early chapters of various books on quantum field theory, such as Wentzel, G. *Quantum Theory of Fields.* New York: Interscience, 1949, and Schweber, S. *Introduction to Relativistic Quantum Field Theory.* Evanston; Row Peterson, 1961.

PROBLEMS

14.1 An infinite frictionless string satisfies the following conditions at $t = 0$:

$$y = Ae^{-bx^2};\ \partial y/\partial t = -2Abvxe^{-bx^2}.$$

Calculate the subsequent motion of the string.

14.2 A string is subjected to the initial conditions:

$$y(x,0) = A\sin(\pi x/S),\ 0 < x < S;\ \partial y/\partial t(x,0) = 0 \text{ everywhere}$$

$$y(x,0) = 0 \text{ elsewhere};$$

Calculate the subsequent motion of the string in two cases:
(a) infinite string;
(b) finite string held at $y = 0$ at $x = 0$, $x = S$.

14.3 Show that Equation 14.49 is expressible as $y(x, t) = f(x - vt) + g(x + vt)$.

14.4 Consider two ways in which an undamped finite string may be set in motion:
(a) $y(x,0) = q(x)$; $y_t(x,0) = 0$;
(b) $y(x,0) = 0$; $y_t(x,0) = \omega_1 q(x)$, where ω_1 is the fundamental frequency of the string.
In which case do the higher harmonics get a larger fraction of the energy?

15

Continuous Systems, Applied Forces, Interacting Systems

Having discussed the behavior of a homogeneous, undamped, undriven string, subject to no external forces except those that hold it fixed at certain points, let us now study the effect of a discontinuity of the string's parameters. We suppose that the string at $x < 0$ has tension $\mathcal{T}$ and mass density μ, and this string is attached at $x = 0$ to another which extends to $x = \infty$, having tension $\mathcal{T}'$, and mass density μ'. There is no practical difficulty in the way of joining strings with different mass densities; however, the difference of tensions will require a force in the x-direction applied at the junction in order to compensate for the net F_x that would otherwise be acting at that point. We do not claim that such a mechanism would be easy to design and apply frictionlessly to the junction of two strings (perhaps a frictionless rod parallel to the y-axis and piercing the string at the junction) but we ask the reader's indulgence because such discontinuities of media commonly affect waves of other types, and we wish to simulate such situations in our study of waves on a string.

In describing the waves on this discontinuous string, we shall find it convenient to express the independent variables in the form $t - x/v, t + x/v$, which forms are equivalent to the $x - vt, x + vt$ used previously. We shall assume that the string on the left ($x < 0$), has a wave $I(t - x/v)$ incident on the junction from $-\infty$, and a reflected wave $R(t + x/v)$ moving toward negative x, and that the string at $x>0$ has only a wave moving away from the junction toward $+\infty$, $T(t - x/V)$. Here clearly v and V are the wave velocities on the two parts of the string:

$$v = (\mathcal{T}/\mu)^{1/2};$$
$$V = (\mathcal{T}'/\mu')^{1/2}. \tag{15.1}$$

The symbols I, R, T stand for "incident," "reflected," "transmitted," respectively. We shall assume that the function I is known in advance, and shall try to discover what reflected and transmitted waves result from the given incident wave.

One condition that must be satisfied is that there is no gap in the string as $x = 0$; at that point, at all times, $I + R = T$:

$$I(t) + R(t) = T(t), \tag{15.2}$$

where x has been set equal to zero in the functions $I(t - x/v)$, $R(t + x/v)$, $T(t - x/V)$.

Also, since the single point $x=0$ has no mass, the y-components of the forces acting on that point from the two sides must be equal and opposite. Taking y_x to be small, we can take the force acting from below to be $\mathcal{T}(-y_x$ at $0-)$ and that from above to be $\mathcal{T}'(y_x$ at $0+)$:

$$\mathcal{T} \frac{\partial}{\partial x} [I + R]_{x=0} = \mathcal{T}' \frac{\partial T}{\partial x} \bigg|_{x=0} \tag{15.3}$$

or

$$-| \mathcal{T}/v | [I'(t) - R'(t)] = -(\mathcal{T}'/V)\, T'(t),$$

or

$$Z[I'(t) - R'(t)] = Z'T'(t),$$

where we have defined

$$Z = (\mu \mathcal{T})^{1/2};$$
$$Z' = (\mu' \mathcal{T}')^{1/2}. \tag{15.4}$$

The primes on the functions I, R, T in Equation 15.3 mean derivatives with respect to the single variables on which they depend; in this case, all the functions depend on the same variable, t. Let us then integrate Equation 15.3 with respect to t from a very negative value at which we can assume the functions all to be zero, to the upper limit which we shall write as t:

$$Z[I(t) - R(t)] = Z'T(t). \tag{15.5}$$

Then, using Equation 15.2 and suppressing the dependence on t, we have

$$Z\,(I - R) = Z'(I + R),$$

or (15.6)

$$R = I\,(Z - Z')/(Z' + Z).$$

Then, from Equation 15.2,

$$T = I\,(2Z)/(Z' + Z). \tag{15.7}$$

With the help of Equation 14.28, we can ask what fraction of the incident energy is reflected and what fraction is transmitted. These two fractions are known as the *reflection* and *transmission coefficients.* Using appropriate subscripts to designate incident, reflected, and transmitted power flow, and recalling that $d/d(x - vt) = -d/vd(t - x/v)$, etc., we have, at $x = 0$,

$$W_I = (\mathcal{T}/v)\,[I'(t)]^2 = ZI'^2,$$
$$W_R = -ZR'^2 = -W_I\,(Z - Z')^2/(Z + Z')^2, \tag{15. 8}$$
$$W_T = Z'T'^2 = W_I\,4ZZ'/(Z + Z')^2.$$

Thus the reflection and transmission coefficients are, respectively,
reflection:

$$(Z - Z')^2/(Z + Z')^2;$$

transmission: (15.9)

$$4ZZ'/(Z + Z')^2.$$

Obviously the sum of these is 1, as energy conservation requires.

For reasons that will shortly become evident, Z and Z' are known as the *impedances* of the respective strings. It is apparent that the way to get maximum energy transfer from the left-hand to the right-hand string is to have $Z = Z'$, that is, to match impedances. In this case there is no reflected wave at all; all the energy gets transmitted, even though the two strings may differ in tension and mass density. What matters is the particular combination of these parameters $Z = (\mu\mathcal{T})^{1/2}$. On the other hand, there is total reflection if one of the impedances equals either zero or infinity—conditions that are most likely to occur through the presence of extreme mass densities. The case $Z' = \infty$ can occur because the string at $x > 0$ has infinite mass density. The total reflection that takes place in this case is the same that arises from a string's being attached to a wall; in each case, the string is attached to an immovable object.

Both Z and Z' being positive, one can see from Equation 15.9 that the transmitted wave T always has the same sign as the incident wave I, at $x = 0$. The reflected wave R may have either sign: if $Z > Z'$, the reflection is uninverted, but if $Z < Z'$ the reflected wave has opposite sign from that of I. The most extreme case of $Z < Z'$ is that discussed above: $Z' = \infty$.

A problem somewhat like that of the joined strings is that of the semi-infinite string acted on by a y-force at its end. This is the limit of a semi-infinite periodic system with input force (or emf), as discussed in Chapter 13. In Chapter 12 we saw that the chain of masses (which goes over into the continuous string in the limit) is the mechanical equivalent of the series-L—shunt-C ladder network. Thus the mechanical input impedance of the chain of masses is given by Equation 13.3, with appropriate changes of electrical to mechanical parameters:

$$Z_i = [(km)(1 - \tfrac{1}{4}\omega^2/\omega_0^2)]^{1/2}, \tag{15.10}$$

where m is the mass per section and k is the effective spring constant, and $\omega_0 = (k/m)^{1/2}$. Writing $k = \mathcal{T}/D$ and $m = \mu D$, we find that $km = \mu\mathcal{T}$, and $\omega_0^2 = k/m = \mathcal{T}/\mu D^2 = v^2/D^2$, which goes to infinity as D as goes to zero. Thus, in the limit, $\omega_0 = \infty$, or the system transmits all frequencies (as we have already seen). In the limit, also, Z_i is a resistance independent of frequency:

$$Z_i = (\mu\mathcal{T})^{1/2}, \tag{15.11}$$

so it can be identified with the Z that we have already used. This identification serves to justify the earlier use of the term "impedance."

Continuing our reference to the results of Chapter 13, we note that $S = Z_1/Z_2$ is now

$$S = -\omega^2/\omega_0^2 \rightarrow 0. \tag{15.12}$$

However, in calculating the limiting expression for $y_t(x)$ at the definite frequency, which is the analog of I_n, we have to remember that n goes to infinity as S goes to zero. So $I_n = I_0E^n$ becomes, at frequency ω,

$$y_t(t) = i\omega y(x) = A[1 - \omega^2/2\omega_0^2 - i(\omega/\omega_0)(1 - \omega^2/4{\omega_0}^2)^{1/2}]^{x/D}, \tag{15.13}$$

where ω_0, as previously given, is v/D, and D is now to be reduced to zero. In the limiting process we need to keep only the constant terms and those linear in D. Thus,

$$i\omega y(x) = A[1 - iD\omega/v]^{x/D} \underset{D\rightarrow 0}{\rightarrow} Ae^{-i\omega x/v}, \tag{15.14}$$

where the relationship, sometimes used as a definition of e^{az}, has been employed:

$$e^{az} = lim_{h\to\infty} [1 + z/h]^{ha}. \tag{15.15}$$

The resulting Equation 15.14 presupposes an $e^{i\omega t}$ time-dependence. Putting this in explicitly gives

$$y(x,t) = [F_0/i\omega\ (\mu\mathcal{T})^{1/2}]\ e^{i\omega(t-x/v)}, \tag{15.16}$$

which is, as it should be, a function of $x-vt$ and therefore represents a wave moving toward $+x$, with velocity v, without changing in any way except in position. Here we have replaced the constant A by the applied force divided by the impedance.

Although we have not discussed it in precisely such terms before, we can now see that the phase velocity of a wave with frequency ω is independent of ω. This property of the system accounts for the dispersion-free propagation of waves, i.e., for the fact that waves like $f(x - vt)$ and $g(x + vt)$ travel at speed v without changing size and shape. In terms of Fourier analysis, we can argue that a disturbance made up of many exponential waves moving in the same direction with the same speed must be the same disturbance, merely displaced, at later times. All the exponential waves $e^{i\omega(t-x/v)}$ move along together, so if they add up to $f(x)$ at $t = 0$ they must add up to $f(x - vt)$ at time t. They do not change their phase relationships; they do not get "out of step."

If the argument about group velocity of Chapter 13 is applied in a situation like the present one, the group velocity $d\omega/dk$ turns out to be identical with the common phase velocity ω/k, where ω and k are the parameters in $e^{i(\omega t-kx)}$. Here we do not need the concept of group velocity which is so useful in systems that propagate waves dispersively; there is only one speed here, v, and everything moves with that speed. It was this property of nondispersive propagation that permitted the use of unspecified functions of $t \pm x/v$, I, R, and T, in the argument about joined semi-infinite strings. In discussing the chain of masses or ladder network we had to treat the case of a sinusoidal disturbance because different frequencies propagate differently and the characteristic impedance is frequency-dependent.

It is necessary to use the same approach in calculating the impedance of a finite or nonuniform string. The case of the joined strings follows rather directly from Equations 13.31 and 13.32. In going to the limit of a continuous system, one has to replace $n\theta$ by $kz = \omega x/v$. Thus, if we wish to know the ratio of incident and reflected amplitudes on one part of a nonuniform string, or wish to know the input impedance at one end of this string, we obtain

$$A/B = e^{-2i\omega x/v}\,(Z_i + Z')/(Z_i - Z'), \qquad (15.17)$$

which is equivalent to Equation 15.6, where x is the position of the discontinuity, and

$$Z_i' = Z_i \frac{Z_i' \sin \omega x/v + iZ' \cos\omega x/v}{Z' \sin\omega x/v + iZ_i \cos\omega x/v}. \qquad (15.18)$$

These expressions are, as stated, specialized in applying only to a sinusoidal excitation, but they are general in that they apply to a string joined at one end to anything that has a calculable impedance Z', not only another string. Anything with impedance not equal to Z_i will produce some reflection and will give Z_i' a periodic dependence on x, the distance to the discontinuity.

In discussing the chain of masses fixed at both ends, we calculated the impedance of the system at the position of any one of the masses. In the present context we can answer similar questions by using the results presented above. One must be careful in combining the impedances of the parts of the system in opposite directions from the point where the total impedance is to be evaluated. If a force is applied to a string at an internal point, the two parts of the string have a common velocity at the point of application, but the forces that they exert on that point add; thus impedance, force/velocity, is the sum of impedances of the two parts. But if an emf is applied between the two sides of the "ladder," say in the middle of one of the series impedances Z_1, the two parts of the network have a common potential difference at the point of application, but the total current flowing from that point is the sum of those going into the two parts of the network. Thus the impedances of the two parts combine as do parallel impedances, as the product over the sum.

Applying the foregoing arguments to a string of length Λ fixed at both ends ($Z' = \infty$), we find its impedance at a point a distance

x from one end:

$$Z(x) = iZ_i \frac{\sin\omega\Lambda/v}{[\sin\omega x/v][\sin \omega(\Lambda - x)/v]}. \tag{15.19}$$

This impedance is purely reactive, there being no way for energy to escape or be dissipated. As in the case of the finite chain of masses (Equations 12.38 and 12.39) the impedance vanishes when ω is one of the normal frequencies of the system, except at the nodes of the standing wave, where Z becomes indeterminate. Physically this behavior means that the system oscillates with no applied force at its normal frequencies (so $F/v = \infty$), except at the nodes, where $v=0$. Such behavior of impedance is a possible criterion for recognizing normal frequencies and wavelengths of normal modes.

The Fourier-integral expansion, hinted at above, is useful in some problems that involve responses to applied forces. If we generalize the argument that led to Equation 14.4″, we can include among the forces acting on dx both a frictional and an applied force, $-ry_t dx$ and $f(x,t)dx$:

$$\mu y_{tt} = \mathscr{T} y_{xx} - ry_t + f(x,t). \tag{15.20}$$

Even without the applied force density (force per unit length) $f(x,t)$, the frictional force invalidates the simple nondispersive solution that we have been studying. Setting $f(x,t) = 0$ for the time being, we can write

$$\mu y_{tt} + ry_t = \mathscr{T} y_{xx}. \tag{15.21}$$

This equation can be attacked by a method known as *separation of variables* which could have been used previously had it been needed. Let us seek solutions of the form

$$y(x,t) = X(x)\, T(t), \tag{15.22}$$

a product of a function of x by a function of t (not to be confused with either kinetic energy or a transmitted wave). Substituting this form into Equation 15.21 and dividing through by the product XT, we get

$$\frac{\mu T'' + rT'}{T} = \frac{\mathcal{T} X''}{X}, \tag{15.23}$$

where the primes denote differentiation of each function with respect to its independent variable. Equation 15.23 states that a function of t only (the left side) equals a function of x only (the right side) at all points x,t in some region. The only way that can be true is that each side equals a constant:

$$\begin{aligned} T'' + 2\gamma T' &= -\upsilon^2 k^2 T; \\ X'' &= -k^2 X. \end{aligned} \tag{15.24}$$

Here υ has its former meaning, $(\mathcal{T}/\mu)^{1/2}$, and $\gamma = r/2\mu$, like the γ of earlier chapters. We have called the constant of separation k^2, a convenient notation. What we have written is equivalent to setting each side of Equation 15.23 equal to $-k^2\mathcal{T}$.

Each of the ordinary differential Equations 15.24 is easy to solve. The general real solution with a given value of $k^2 > \gamma^2/\upsilon^2$ is

$$y(x,t) = e^{-\gamma t}\,[A\cos(\omega' t + kx + \theta_1) + B\cos(\omega' t - kx + \theta_2)], \tag{15.25}$$

where $\omega'^2 = k^2\upsilon^2 - \gamma^2$. If the motion is overdamped ($k^2 < \gamma^2/\upsilon^2$), the general real solution is, for a given value of k^2,

$$y(x,t) = A\cos(kx + \theta_1)\,e^{-b_1 t} + B\cos(kx + \theta_2)\,e^{-b_2 t}, \tag{15.26}$$

where

$$\begin{aligned} b_1 &= \gamma + \sqrt{\gamma^2 - \upsilon^2 k^2}; \\ b_2 &= \gamma - \sqrt{\gamma^2 - \upsilon^2 k^2}. \end{aligned} \tag{15.27}$$

The case of critical damping, $v^2k^2 = \gamma^2$, can be written also, as in Chapter 2.

If the damped string is fixed at $x=0$ and $x=\Lambda$, then the cosine functions of k must be sine functions with

$$k_n = n\pi/\Lambda, \tag{15.28}$$

as in Chapter 14. If the damping, γ, is large enough for some of the k_n to be less than γ/v, these modes will not oscillate at all, but will move as Equation 15.26 indicates. But if n is high enough—perhaps even if it is only 1—the motion will be damped-harmonic, like that of a freely moving damped oscillator as studied in Chapter 2. The frequencies are all lowered by the damping from vk_n to $(v^2k_n^2 - \gamma^2)^{1/2}$, so they no longer are quite equally spaced and, even without the factor $e^{-\gamma t}$, would not in general add up to periodic motion. The relationship that led to nondispersive propagation on the undamped string, $\omega = kv$, is no longer valid here, so propagation is dispersive in general.

Whether or not there is damping, the method of separation of variables can be expected to yield very specialized particular solutions of the partial differential equation. Thanks to the linearity of the equation, however, such particular solutions with various values of the separation constant can be added together to produce more general solutions.

Now let us return to the question of how a string responds to a given applied force. In order to avoid such complications as branch points, we shall make a slightly unrealistic assumption which we shall later see how to interpret: in addition to the $-y_{xx}$ restoring force and the $-ry_t$ frictional force and the applied force $f(x,t)$, we shall suppose that there is an additional restoring force per unit length equal to $-h\gamma^2\mu y$, where h is a real nonnegative number, and $\gamma = r/2\mu$ as previously defined. The equation of motion is now

$$my_{tt} - \mathcal{T}y_{xx} + ry_t + h\gamma^2\mu y = f(x,t), \tag{15.29}$$

where f is the applied fore per unit length of the string. Now let us assume that y and f can be expressed as (double) Fourier integrals:

$$y(x,t) = (2\pi)^{-1} \iint \beta(k,\omega)\, e^{i(\omega t - kx)}\, d\omega dk;$$

$$f(x,t) = (2\pi)^{-1} \iint \alpha(k,\omega)\, e^{i(\omega t - kx)}\, d\omega dk;$$

whence (15.30)

$$\alpha(k,\omega) = (2\pi)^{-1} \iint f(x',t') e^{-i(\omega t' - kx')}\, dx'\, dt'.$$

In assuming the Fourier integral form for y we are tacitly assuming that no other force than f acts, or that any motion due to earlier forces has been damped out. Thus we are seeking the response of the string to f.

Substituting from Equation 15.30 into 15.29, we find that

$$\beta(k,\, \omega) = \frac{\alpha(k,\omega)}{\mathcal{T}k^2 - \mu\omega^2 + i\omega r + h\mu\gamma^2}, \tag{15.31}$$

so Equation 15.30 now gives

$$y(x,t) = \frac{1}{2\pi} \iint \frac{\alpha(k,\omega) e^{i(\omega t - kx)} d\omega dk}{\mathcal{T}k^2 - \mu\omega^2 + ir\omega + h\mu\gamma^2}, \tag{15.32}$$

Knowing the applied force density $f(x,t)$, we can calculate $\gamma(k,\omega)$, which we can then use in Equation 15.32 to get the response to f.

It is possible also, as in Chapter 6, to substitute for $\alpha(k,\omega)$ in Equation 15.32 from Equation 15.30, then reverse the order of the two (double) integrations and thus get a Green's-function expression for y:

$$y(x,t) = \iint G(x - x',t - t')\, f(x',t')\, dx'dt', \tag{15.33}$$

where

$$G(x - x',t - t') = (2\pi)^{-2} \iint \frac{e^{i[\omega(t-t') - k(x-x')]} dk d\omega}{\mathcal{T}k^2 - \mu\omega^2 + ir\omega + h\mu\gamma^2}. \tag{15.34}$$

We shall do the k-integration first, so we shall express the denominator in suitable form:

$$G = \frac{1}{4\pi^2 \mathscr{T}} \iint \frac{e^{i[\omega(t-t')-k(x-x')]}\, dk\, d\omega}{(k - k_1)(k - k_2)}, \tag{15.35}$$

where

$$\begin{aligned} k_1 &= \pm \frac{1}{v} \sqrt{\omega^2 - 2i\gamma\omega - h\gamma^2}; \\ k_2 &= \mp \frac{1}{v} \sqrt{\omega^2 - 2i\gamma\omega - h\gamma^2}. \end{aligned} \tag{15.36}$$

The two roots k_1 and k_2 being negatives of each other, one lies in the upper and the other in the lower half-plane. We shall define k_1 to be the one in the upper half-plane. Thus, closing the contour by means of an infinite semicircle S_+ or S_-, we choose S_+ and thus encircle k_1 when $x < x'$ (note $-$ sign before k in the exponent), and we encircle k_2 in the negative direction when $x > x'$. In each case we get a residue in which k_1 or k_2 has been substituted for k in the analytic part of the integrand.

The procedure up to this point presents no particular difficulties. The problems begin after the calculation of residues from the k-integration, when we undertake to do the ω-integration. The substitution of square-root functions of ω for k leads to the integrand now being a double-valued function of ω, whereby the integration becomes a rather different problem from those that we have previously encountered in the complex plane.[1] In order to avoid having to embark on new mathematical subject matter, we shall deal only with the cases in which k_1 and k_2 are rational functions, i.e., in which their radicands are perfect squares. There are two such cases that are not trivial: $h = 1$ and $\gamma = 0$. Let us look first at the case $h = 1$:

$$\begin{aligned} k_1 &= -(\omega - i\gamma)/v; \\ k_2 &= +(\omega - i\gamma)/v. \end{aligned} \tag{15.37}$$

Now the ω -integrals are:

$$G = \frac{-1}{4\pi Z}\int \frac{e^{i\{\omega(t-t')-[(\omega-i\gamma)/v](x'-x)\}}}{\omega-i\gamma}\,d\omega \text{ (for } x < x');$$

(15.38)

$$G = \frac{-i}{4\pi Z}\int \frac{e^{i\{\omega(t-t')-[(\omega-i\gamma)/v](x-x')\}}}{\omega-i\gamma}\,d\omega \text{ (for } x > x').$$

In both these integrals one has the choice of closing the contour via S_+ or S_-, and in each one uses the integral on S_- (which is zero) when the coefficient of ω in the exponent is negative, thus getting zero for G. When the coefficient of ω in the exponent is positive, one uses S_+, getting G equal to the integral around C_+, that is, $2\pi i$ times the residue at $\omega = i\gamma$:

$$G(x - x', t - t') = e^{-\gamma(t-t')}/2Z$$

(for $x < x'$ and $t - t' > (x' - x)/v$);

$$G(x - x', t - t') = e^{-\gamma(t-t')}/2Z \tag{15.39}$$

(for $x > x'$ and $t - t' > (x - x')/v$);

$$G(x - x', t - t') = 0$$

(for $t - t' < | x - x' | /v$).

Thus the Green's function is a pair of step functions of position, with a damping factor, propagating out from a point where force is applied, in both directions, with speed v.

If $\gamma = 0$, both the last two terms on the left side of Equation 15.29 are absent, and the equation describes the frictionless string. The Green's function is changed solely by losing the damping factor; it consists of two step functions moving apart, each going at speed v.

When we carried out the limiting process on the chain of masses, it was obvious that the limiting case should be interpreted as a continuous string under tension. Although we have mentioned the limiting form of the corresponding electrical system, the series-L–shunt-C ladder network, we have not suggested a physical interpretation for it. In view of the fact that a pair of straight parallel wires has a certain series inductance per unit length and a certain capacitance per unit length between the wires, it seems reasonable to interpret the limiting network to be simply a parallel-wire transmission line.[2] The analog of μ, mass per unit length, is L, inductance per unit length, whereas $\mathcal{T}$, tension, corresponds to $1/C$, the reciprocal of capacitance per unit length. Thus velocity is, in the limit.

$$v = (LC)^{-1/2}, \tag{15.40}$$

which always turns out to be c, the speed of light, despite the fact that L and C separately depend on the ratio of diameter to separation of the two wires. The phase velocity of c agrees with the results of a careful calculation by means of electromagnetic theory in which the two wires are assumed to have no resistance (infinite conductivity); such a calculation is preferable to the use of lumped-parameter circuit theory (what we have been using) to simulate a system with distributed parameters. The problem is, aside from radiation, that electric and magnetic fields from one part of a wire affect the same wire and the other wire at other points and other times; one can no longer properly assume that impedances are localized. So we must conclude that the limiting form of the ladder network is not too bad, but not so realistic as the limiting form of the chain of masses.

If we inquire about the limiting forms of other ladder networks as the sections become infinitely small, we find that in general these limiting forms make no sense at all; they cannot be even approximately constructed. If the series and shunt impedances cannot be construed as impedances or admittances per unit length in a continuous system, there is no limiting form. For example, the series-C—shunt-L network makes no sense in the limit; the series capacitances have higher, not lower, impedance for shorter lengths, and the shunt inductances give rise to similar problems. If one wished to construct this system in the limiting case of a continuous system, one would have to perform, among other miracles, the feat of building an infinite series capacitance into each infinitesimal length of the system.

NOTES

1. When $h = 0$, the ω −integral for $t > t'$ reduces to an integral around a pair of branch points connected by a *cut*, i.e., a segment that one may not cross. This integral in turn reduces to the integral along this segment of the difference between the values assumed by the integrand on the two sides of the cut. This integral involves functions that we do not have time and space to deal with here. On the general subject of integrals of multiple-valued functions, see the works previously cited by Churchill, R.V., J.W. Brown, and R.F. Verhey. *Complex Variables and Applications.* 3d ed. New York: McGraw Hill, 1974; also Franklin, P. *Functions of Complex Variables.* Englewood Cliffs, NJ: Prentice-Hall, 1947; and Smith, L.P. *Mathematical Methods for Scientists and Engineers.* Englewood Cliffs, NJ: Prentice Hall, Dover Reprint, 1961.
2. For a calculation of L and C per unit length of a parallel-wire line, and also a calculation of wave propagation guided by such wires, see: Jackson, J.D., *Classical Electrodynamics.* 2d ed. New York: Wiley, 1975.

PROBLEMS

15.1 A finite undamped string fixed at its ends is set in motion by the force density $f(x,t) = \delta(t)\,\delta(x - x_0)$. Calculate the resulting energy content of each standing wave and show its relationship to that wave's amplitude at $x = x_0$.

15.2 The string in Problem 15.1 is set in motion by $f(x,t) = \delta(t)[\delta(x - \frac{1}{4}\Lambda) - \delta(x - \frac{3}{4}\Lambda)]$. Show that none of the odd modes is present thereafter.

15.3 An infinite homogeneous string has a point mass attached to it at $x = 0$. Calculate the reflected and transmitted waves resulting from an incident wave $I\cos(\omega t - kx)$ on the part of the string where $x < 0$.

15.4 A finite undamped string fixed at its ends is driven by a force density $f(x,t) = A\cos\omega t\,\delta(x - x_0)$. Calculate the motion of the string at the frequency ω, assuming that any transient effects have died out despite the absence of friction.

15.5 The ends of three strings are fastened together at the origin. The strings all extend to infinity in different directions in the zx-plane, and are constrained to vibrate only in the y-direction. They all have different mass densities, and their tensions are adjusted to produce equilibrium at the point of joining. Determine how the mass densities, tensions, and directions of the strings must be related so that a wave approaching the junction along one of the strings will *not* be reflected at all. If this condition is satisfied, can a wave coming in on one of the other strings escape reflection?

15.6 Two infinite strings cross at right angles and are joined where they cross. They have different mass densities and tensions. Both are constrained to vibrate perpendicular to their common plane. If a wave comes in to the junction along one of the strings, what fraction of its energy goes out along each of the four possible paths?

15.7 Calculate the normal modes and frequencies of a finite string fixed at both ends, with a point mass attached to its center. Note that the Lagrangian of this system is the sum of Lagrangians of string and mass, where the latter contains a term representing the potential energy of the mass subject to the force exerted on it by the string (assumed to be solely in the y-direction).

15.8 Consider the following possibilities for the Z' in Equations 15.17 and 15.18, verifying in each case that the equations give reasonable results: $Z' = Z_i$, $Z' = 0$, $Z' = \infty$, Z' = pure reactance.

15.9 What are the normal frequencies of a string that is under tension, but both ends of which can move freely? of a string fixed at one end and free to move at the other?

15.10 Derive an equation for the normal frequencies of two strings joined end to end with their opposite ends fixed, assuming that they may have different mass densities, tensions, and lengths.

16

Other Linear Problems

We have now devoted fourteen chapters—Chapter 8 was a digression—to a few rather simple systems that can be quite thoroughly analyzed by linear mathematical techniques. The systems treated were deliberately chosen to be simple, so that the methods of analysis would be as clear as possible. Now, in order to convey a better idea of the variety of systems to which these methods are applicable, we look briefly at some other phenomena that are describable by linear equations.

The first of these is the propagation of a plane sound wave through a tube of cross-sectional area A, lying along the x-axis. Let us take the density of the gas in the tube in the absence of sound to be ρ_0 and the pressure to be p_0; when a sound wave is present, the density will be $\rho_0 + \rho'(x,t)$ and the pressure will be $p_0 + p'(x,t)$. Now let us imagine two planes perpendicular to the x-axis, at x and at $x+\Delta x$, and let us so define them that they move with the gas, always having the same gas between them. Then, letting h be the displacement of the gas from its equilibrium position, we can apply Newton's second law to the gas between the two planes:

$$\rho A(\Delta x + \Delta h)\ddot{h} = F_x = -(\partial p/\partial x)(\Delta x+\Delta h)\, A,$$

or (16.1)

$$\rho\ddot{h} = -\partial p/\partial x = -(dp/d\rho)(\partial\rho/\partial x).$$

The left side of the top equation is the time derivative of the x-momentum of the constant mass of gas between the two moving planes, and the right side is the force acting on this mass by virtue of the pressure gradient which produces unequal forces on the two equal areas A. In the second equation this area and the thickness of the slab of gas, $\Delta x + \Delta h$, have been divided out and, on the assumption that p can be expressed as a function of density ρ alone, the latter has been substituted for p in the gradient.

Let us now express explicitly the fact that we have used already, the constancy of the mass of gas between the two planes:

$$A\rho_0\Delta x = A(\rho_0 + \rho')(\Delta x + \Delta h),$$

or (16.2)

$$0 = \rho_0\Delta h + \rho'\Delta x + \rho'\Delta h.$$

Now dropping the last term, which is the product of small quantities, and reinterpreting the other terms, we have

$$\frac{\rho'}{\rho_0} = -\frac{\Delta h}{\Delta x} = -\frac{\partial h}{\partial x}. \tag{16.3}$$

Now Equation 16.1 contains $\partial\rho/\partial x$, which is the same as $\partial\rho'/\partial x$ (ρ_0 being constant). We can substitute for this derivative from the x-derivative of Equation 16.3:

$$\rho\frac{\partial^2 h}{\partial t^2} = -\frac{\partial p}{\partial\rho}\left(-\rho_0\frac{\partial^2 h}{\partial x^2}\right). \tag{16.4}$$

Introducting the bulk modulus of the gas, $\rho dp/d\rho = B$, we have

$$h_{tt} = (B/\rho)h_{xx}, \quad (16.5)$$

a wave equation like the one for the string, with phase velocity

$$v = (B/\rho)^{1/2}. \quad (16.6)$$

Newton went through reasoning like that just presented and then used the ideal gas equation, which can be written

$$p = CRT\rho, \quad (16.7)$$

to get

$$v^2 = B/\rho = dp/d\rho = CRT, \quad (16.8)$$

where the constant C is now known to be the reciprocal of the molecular weight of the gas, provided the other quantities are expressed in cgs units. One of Newton's frustrations was the inaccuracy of this expression for sound velocity; it gave numbers that were consistently too low by about 20 to 30 percent.

It is known now that $dp/d\rho$ should not be calculated, as Newton did it, on the assumption that the gas is isothermal; the changes occurring during the passage of a sound wave—at least in the audible range of frequencies—are much more nearly adiabatic than isothermal. An ideal gas that is changing adiabatically satisfies

$$p = \text{constant } \rho^{\gamma}, \quad (16.9)$$

whence

$$v^2 = dp/d\rho = \gamma p/\rho = \gamma CRT;$$

That is, the calculated v^2 is increased by the factor γ, which here means the ratio of specific heats of the gas, c_p/c_v. This prediction is well borne out by available data.

If we look critically at the foregoing derivation that led to Equation 16.5, we can see various approximations. One of these is hidden in the derivation of Equation 16.1, where it is assumed

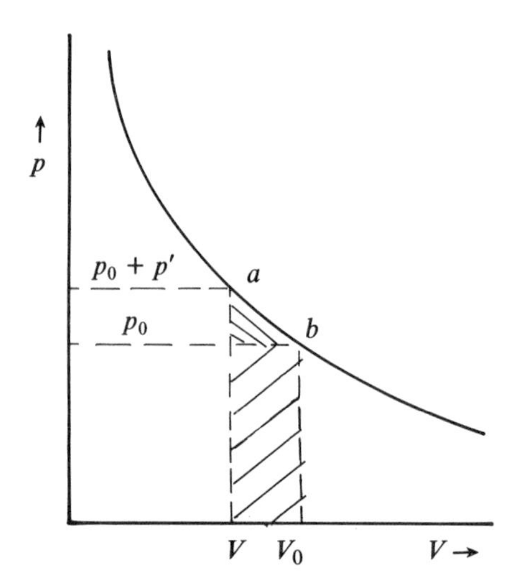

Figure 16.1. Vp diagram showing the effect on the state of a gas of a weak, non-dissipative sound wave. Note that p and V change in both directions along the same curve (normally the curve of adiabatic changes), so the wave does no net work on the gas.

that the value of x which is the undisplaced position of the gas whose momentum change enters the equation is the same value at which the right side should be evaluated. This assumption is valid only for small displacements h. Also, treating $dp/d\rho$ as a constant presupposes small amplitudes of vibration, for if ρ' is comparable to ρ_0, or even considerably smaller, the bulk modulus will change enough during the passage of the sound wave to introduce nonlinearities into the equations. Wave equations for waves of large amplitude, like the DE for large pendulum or oscillator amplitudes, are nonlinear; so is the string equation in such cases. The most strikingly nonlinear sound-wave phenomenon is a shock wave, which travels faster than sound of small amplitude and spontaneously develops a discontinuous pressure rise at the leading edge.

We end the discussion by calculating the potential-energy density of a plane sound wave, assuming that its amplitude is small and that its propagation is nondissipative. The work done in changing a volume of gas from V to $V + dV$ at pressure p is $dW = -pdV$. This is work done *on* the gas; it can be recovered when the volume returns to its original value, if the gas follows the same curve in the pV-plane both times. Let us suppose that it does, and that the curve followed is the one in Figure 16.1. In the figure we have marked p_0 and V_0, the pressure and the molar volume of the undisturbed gas, and we have marked also certain altered pressure and volume values that might be attained during the passage of a sound wave. (Note that, typically, $p_0 = 10^6$ dyne/cm^2 and $p' = 10^{-3}$ dyne/cm^2, so the fractional changes being contemplated are very small, for example, 1 part in 10^9).

The shaded area in the figure thus represents the work done on the gas in the change from state b to state a along the curve. Treating the small piece at the top as a triangle, we have, for one mole of gas,

$$W = p_0(V_0 - V) + \tfrac{1}{2}p'(V_0 - V). \qquad (16.10)$$

We wish to know the work per unit volume rather than per mole, so we shall divide Equation 16.10 by V_0. Then each term contains $(V_0 - V)/V_0$, which is, for small fractional changes, the same as p'/p_0. Furthermore, the pressure excursion p' can be written as

$$p' = \rho'(dp/d\rho) = B\,\rho'/\rho_0. \qquad (16.11)$$

Also, by equation 16.3,

$$\rho'/\rho_0 = -\partial h/\partial x. \qquad (16.12)$$

Thus we can write the work done on a unit volume of gas as

$$w = -p_0\,\partial h/\partial x + \tfrac{1}{2}B(\partial h/\partial x)^2, \qquad (16.13)$$

which is thus the potential energy per unit volume for a plane sound wave propagating along the x-axis. If we imagine integrating this quantity with respect to x in order to get the total PE of a sound wave at a given instant, we find that the first term makes a contribution that becomes vanishingly small compared to that of the second term as the range of integration gets large—because $\partial h/\partial x$ integrates to h, which never gets very big, whereas the second term is positive-definite, so its integral increases without limit. Thus we can omit the first term from our definition of PE density, getting

$$u = \tfrac{1}{2}Bh_x^2. \tag{16.14}$$

Then the Lagrangian density is

$$\mathscr{L} = \tfrac{1}{2}\rho h_t^{\,2} - \tfrac{1}{2}Bh_x^2. \tag{16.15}$$

It can be illuminating to describe a wave of the form $f(x - v_0 t)$ in a frame of reference in which the wave is at rest. The medium (string, gas, etc.) is then flowing through a stationary wave; the velocity of the medium at infinity is $-v_0$ if the wave is localized, and other quantities likewise assume fixed values at infinity. Using this mode of description of a localized sound wave, we use the same two physical principles as before, Newton's second law and conservation of mass. The two planes moving with the gas, with separation Δx where the density is ρ, have separation Δx_0 at infinity where the density is ρ_0. Conservation of mass requires that

$$\rho\Delta x = \rho_0 \Delta x_0. \tag{16.16}$$

The same physical fact has a different expression in the requirement that the fluid flow in gm/cm^2-sec is independent of x:

$$\rho v = \rho_0 v_0. \tag{16.17}$$

The mass in the interval Δx is $\rho A \Delta x$, and the net force acting on that mass is $-(\Delta x) A\,dp/dx$, so

$$-A\Delta x\,\frac{dp}{dx} = \rho A \Delta x\,\frac{dv}{dt}. \tag{16.18}$$

The time derivative in this equation is a so-called "convective" derivative, i.e., one that is evaluated at a point moving with the gas. In three dimensions such a derivative operator is

$$\frac{d}{dt} = \frac{\partial}{\partial t} + v \cdot \nabla. \qquad (16.19)$$

In our case the partial derivative is zero because the flow is steady, so Equation 16.18 becomes

$$-\frac{\rho_0 \Delta x_0}{\rho}\frac{dp}{dx} = \rho_0 \Delta x_0\, v \frac{dv}{dx}, \qquad (16.20)$$

or

$$\frac{dp}{dx} + \rho_0 v_0 \frac{dv}{dx} = 0, \qquad (16.21)$$

so

$$p + p_0\, v_0 v = \text{const.} = p_0 + \rho_0 {v_0}^2. \qquad (16.22)$$

Here we have used Equations 16.16 and 16.17. Now, rearranging, using Equation 16.17 again, and using two terms of a Taylor series to express p as a function of ρ, we have

$$(\rho - \rho_0) \left.\frac{dp}{d\rho}\right|_0 \cong = \rho_0 v_0^2 \left(1 - \frac{\rho_0}{\rho}\right),$$

or $$(16.23)$$

$$v_0^2 = \frac{\rho}{\rho_0} \left.\frac{dp}{d\rho}\right|_0 = \frac{B_0}{\rho_0}$$

This is the same expression for sound speed that is given in Equation 16.6. Possibly this derivation offers new insights into the assumptions that underlie this result.

If one treats the infinite string by a similar argument, one contemplates a string moving at speed v_0 around localized stationary curves. If the string has mass μ per unit length, each point on the string must be subject to a centripetal force $\mu {v_0}^2/R$, where R is the radius of curvature of the string at the point. This centripetal force is produced by the tension in the string, assumed constant, and is directed toward the center of curvature at the point. Figure 16.2 shows some of the relationships.

The figure shows that the force on an infinitesimal piece of a string is radial, and that it is the sum of contributions from the two ends; the force on ds is

$$dF = 2\mathcal{T}\sin(d\theta/2) = \mathcal{T}d\theta = \mathcal{T}ds/R. \qquad (16.24)$$

Equating this to the centripetal force needed to keep the string moving in its curve, we get

$$\mu ds\, v_0^2/R = \mathcal{T}ds/R,$$

or

$$v_0^2 = \mathcal{T}/\mu, \qquad (11.25)$$

as previously argued. This argument makes it evident that the angle θ does not have to be small; all one has to assume is that tension is constant along the length of the string. This assumption probably is seriously inaccurate when amplitudes of motion are large. If tension is not constant along the string, there will be longitudinal as well as lateral accelerations, so various modes of

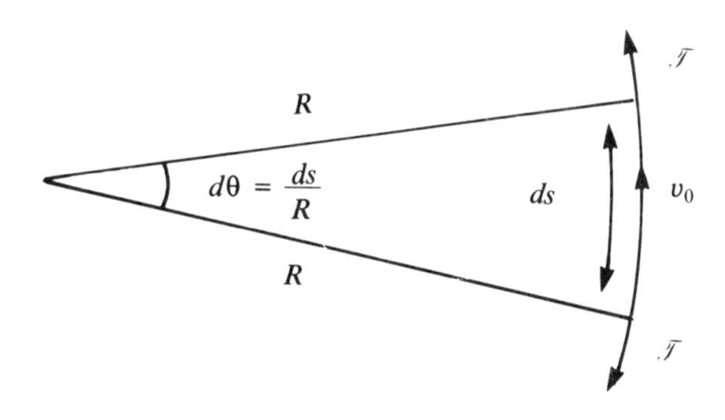

Figure 16.2. A piece of string moving at speed v_0 around a stationary curve.

motion will be simultaneously present and will interact with each other.

A third type of wave that we shall describe in the rest frame of the wave is a small wave on the surface of deep water. If D is the depth of the water at infinity and $D+h$ is its depth where a wave has height h [$h = h(x)$], $D \gg h$ makes it reasonable to assume that the velocity v of the water is uniform from the bottom to the surface as it flows under any given point of the stationary surface wave. Then conservation of water can be expressed as

$$v/v_0 = D/(D + h),$$

or (16.26)

$$v^2 - v_0^2 = v_0^2[-2h/(D + h) + h^2/(D + h)^2].$$

The same quantity can be expressed by means of Bernoulli's principle, which applies to stationary flow. At any given distance z above the bottom,

$$\tfrac{1}{2}v^2 + g(D - z + h) = \tfrac{1}{2}v_0^2 + g(D - z),$$

or (16.27)

$$v^2 - v_0^2 = -2gh.$$

Combining these results, we have

$$gh = v_0^2\,[h/(D + h)[1 - h/2(D + h)],$$

or (16.28)

$$v_0^2 \simeq Dg,$$

where we have neglected h in comparison with D. Note that v in Equation 16.26 means the horizontal component of velocity, whereas that in Equation 16.27 is the total speed. Furthermore, $\mathbf{v}$ at the bottom is precisely horizontal, but $\mathbf{v}$ at the surface is tangent to the surface, following the shape of the wave. We can safely pretend that **v** is horizontal everywhere (as we have done, in effect) only if the surface is practically horizontal. This condition, made more quantitative, can be stated as a requirement that all the wavelengths present in the spectrum of the wave are long compared to the depth, D.

These three derivations in the rest frames of waves are all based on the assumption that nondispersive propagation is possible, so it makes sense to assume the existence of a wave that *has* a rest frame. A medium in which all waves are damped would obviously be unsuited to such analysis. Once this assumption is made, our three examples have the common property that any function of x, provided it is not too large, qualifies as a wave displacement function; as for the string, also in the other cases $f(x - v_0 t)$ is a possible wave, whatever (sufficiently differentiable) function f may be. A medium like the chain of masses lacks this property; on account of dispersion, only sinusoidal functions can play the role of f. Media with nonlinear properties might also permit only certain functions f, if any.

The periodic and continuous systems studied so far in this book have all been one-dimensional. However, most phenomena can be adequately described only in three-dimensional space. In the rest of this chapter we look briefly at some three-dimensional phenomena, starting with some whose partial differential equations are of first (instead of second) order in time. Such equations, still linear, are used in one dimension and in two and three dimensions. We shall here indicate the derivation of two of them in three dimensions and then briefly discuss some methods of solution. Let us consider a three-dimensional homogeneous substance in which heat flows by conduction at a rate proportional to the negative gradient of temperature:

$$\mathbf{J} = -k\nabla T, \qquad (16.29)$$

where $\mathbf{J}$ is heat flow per unit area per second, a vector pointing in the direction in which heat is flowing, and k is a constant known as *thermal conductivity*. From Equation 16.29 we can derive the divergence of $\mathbf{J}$:

$$\nabla \cdot \mathbf{J} = -k\nabla^2 T, \qquad (16.30)$$

where ∇^2, the Laplacian (compare Equation 5.10), is the divergence of the gradient. In Cartesian coordinates,

$$\nabla^2 T \equiv T_{xx} + T_{yy} + T_{zz}. \tag{16.31}$$

Now Gauss' theorem (or divergence theorem) states that the integral of the divergence of a vector function over a volume equals the flux of the vector outward through the surface surrounding the volume; this is a mathematical result that follows from partial integration. Thus by integrating Equation 16.30 over a volume V we get the rate at which heat is flowing out of V. This has to equal the rate at which it is being produced in V, minus the rate at which it is accumulating within V. The latter rate can be expressed in terms of a specific heat c per unit volume times a rate of change of temperature, integrated over V. Thus

$$-k\int_V \nabla^2 T\,dV = \int_V \nabla\cdot\mathbf{J}\,dV = \oint_S \mathbf{J}\cdot\mathbf{dS} = \int_V [f(\mathbf{r},t) - c\partial T/\partial t]\,dV. \tag{16.32}$$

This equation is valid for any volume V that one may choose. Thus, it is valid as a relationship among the integrands at each point:

$$k\nabla^2 T - c\partial T/\partial t = -f(\mathbf{r},t). \tag{16.33}$$

Here $f(\mathbf{r},t)$ is the rate of production of heat, per unit volume per sec, at point $\mathbf{r} = (x,y,z)$ and time t.

Another physical phenomenon described by the same sort of PDE is diffusion in a homogeneous medium. Let ϕ be the concentration in gm/cm^3 of a substance (solute) that is dissolved in another (solvent). If this concentration ϕ varies from point to point, there will be diffusion of the solute within the solvent. The flow rate of solute by diffusion in gm/cm^2 sec is often given by

$$\mathbf{J} = -D\nabla\phi, \tag{16.34}$$

where D is a constant known as the *coefficient of diffusion.* By a conservation argument for the solute like the one applied to heat above, we can with the aid of Gauss' theorem arrive at

$$D\nabla^2\phi - \partial\phi/\partial t = -f(\mathbf{r},t), \tag{16.35}$$

where now $f(\mathbf{r},t)$ is the production-rate density of the solvent, in gm/cm^3 sec, by chemical processes such as metabolism.

Another equation with the same form, except that t is replaced by (it) is the Schrödinger equation of quantum theory for a force-free particle in three dimensions. Without pretending to derive it, we simply exhibit it:

$$\nabla^2\psi - (2m/\hbar)\partial\psi/\partial\,(\mathrm{it}) = 0. \tag{16.36}$$

Here ψ is known as the *wave function* of a particle of mass m; $\hbar$ is Planck's constant divided by 2π. There is no "source" term here like the $f(\mathbf{r},t)$ of Equations 16.33 and 16.35.

If we wish to apply any of the foregoing equations in a situation in which nothing varies in the y and z directions (either because the geometry is one-dimensional as in a long tube, or because the quantities of interest are constant over large areas perpendicular to the x-axis) the Laplacian operator reduces to the second derivative with respect to x, and we have a one-dimensional PDE somewhat like the string wave equation (except that $\partial/\partial t$ is present instead of $\partial^2/\partial t^2$). Conversely, the extension of a wave equation of second order in t, like that of the string, to three dimensions, merely entails the replacement of $\partial^2/\partial x^2$ by the Laplacian.

We shall study an equation like Equations 16.33, 16.35, and 16.36 in order to derive a Green's function and discover two of its uses. We take the equation to be

$$a\nabla^2\phi - b\phi_t = -f(\mathbf{r},t). \tag{16.37}$$

We assume that ϕ and f are expressible as multiple Fourier integrals:

$$\phi(\mathbf{r},t) = \frac{1}{(2\pi)^2}\int \beta(\mathbf{k},\omega)e^{i(\omega t-\mathbf{k}\cdot\mathbf{r})}\,d^3k\,d\omega \tag{16.38}$$

$$f(\mathbf{r},t) = \frac{1}{(2\pi)^2}\int \alpha(\mathbf{k},\omega)e^{i(\omega t-\mathbf{k}\cdot\mathbf{r})}d^3k\,d\omega$$

$$(16.39)$$

$$\alpha(\mathbf{k},\omega) = \frac{1}{(2\pi)^2}\int f(\mathbf{r}',t')e^{-i(\omega t'-\mathbf{k}\cdot\mathbf{r}')}d^3x'dt'$$

Here d^3x' means $dx'dy'dz'$, d^3k means $dk_xdk_ydk_z$, and $\mathbf{k}\cdot\mathbf{r}$ means $xk_x+yk_y+zk_z$. Substituting from Equations 16.38 and 16.39 into 16.37, we find that

$$\beta(\mathbf{k},\omega) = \frac{\alpha(\mathbf{k},\omega)}{\alpha k^2 + i\omega b} = \frac{\alpha/ib}{\omega - iek^2/b}, \tag{16.40}$$

where $k^2 = k_x^2 + k_y^2 + \mathrm{k}_z^2$. Thus

$$\phi(\mathbf{r},t\ ) = \frac{-i}{(2\pi)^2 b}\int \frac{\alpha(\mathbf{k},\omega)e^{i(\omega t-\mathbf{k}\cdot\mathbf{r}')}d^3k\,d\omega}{\omega - iak^2/b} \tag{16.41}$$

is the solution of Equation 16.37 that vanishes when f does; that is, the response of the system to f. Now, as we have done before, we substitute for α in Equation 16.41 from Equation 16.39, and reverse the order of integrations, getting

$$\phi(\mathbf{r},t) = \int G(\mathbf{r}-r',t-t')\, f(\mathbf{r}',t')\ d^3x'dt', \tag{16.42}$$

where

$$G(\mathbf{R},T) = \frac{-i}{(2\pi)^4 b}\int \frac{e^{i(\omega T-\mathbf{k}\cdot\mathbf{R})}d^3k\,d\omega}{\omega - iak^2/b} \tag{16.43}$$

Now, doing the ω integration first, we note that the only pole is in the upper half-plane and therefore contributes only when $T > 0$, that is, when $t > t'$. So $G = 0$ when $t < t'$, and when $t > t'$,

$$G(\mathbf{R},T) = \frac{1}{(2\pi)^3 b}\int e^{-ak^2T/b-i\mathbf{k}\cdot\mathbf{R}}d^3k. \tag{16.44}$$

This integrand is a product of functions of the three components of **k**, so the integral is a product. Any one of the three factors is a one-dimensional Green's function; such a factor can be evaluated by the method shown in Equations 4.38–4.41. Then, multiplying the three factors together, we get, for $t > t'$,

$$G(\mathbf{r} - \mathbf{r}', t - t') = \sqrt{b}[4\pi a(t - t')]^{-3/2} e^{-b(\mathbf{r}-\mathbf{r}')^2/4a(t-t')}. \tag{16.45}$$

Like other Green's functions, this one can be interpreted as the response of the system to a delta-function source. If

$$\begin{aligned} f(\mathbf{r}'',t'') &= \delta^3(\mathbf{r}'' - \mathbf{r}')\delta(t'' - t') \\ &= \delta(x'' - x')\delta(y'' - y')\delta(z'' - z')\delta(t''-t'), \end{aligned} \tag{16.46}$$

an instantaneous point source at $\mathbf{r}', t'$, the resulting behavior of ϕ is simply the G of Equation 16.45. G is the solution of

$$a\nabla^2 G - b\frac{\partial G}{\partial t} = -\delta^3(\mathbf{r} - \mathbf{r}'')\delta(t - t') \tag{16.47}$$

that vanishes when $t < t'$.

The same function G for an equation like Equation 16.37 has a different interpretation, also, as suggested in Equations 14.52 and 14.53. There the calculation diverged, but here we can carry through the determination of ϕ from initial conditions. Let us take $f(\mathbf{r},t) = 0$ in Equation 16.37, and use the fact that the solution of an equation of first order in t is determined by specification of the solution over all space at one time. First we substitute the form in Equation 16.38 into Equation 16.37, finding that

$$ak^2 + i\omega b = 0. \tag{16.48}$$

Thus in the integral of Equation 16.38 ω is no longer independent of **k**. The quadruple integral reduces to a triple one:

$$\phi(\mathbf{r},t) = (2\pi)^{-3/2} \int g(\mathbf{k})\, e^{-ak^2t/b - i\mathbf{k}\cdot\mathbf{r}}\, d^3k. \tag{16.49}$$

This is the general solution of the homogeneous equation, Equation 16.37, with $f = 0$; like the general solution of the string equation, it involves an arbitrary function, here $g(\mathbf{k})$. Now we evaluate this function by means of initial conditions. We take as given the dependence of ϕ on $\mathbf{r}$ at a time t', and express it as

$$\phi(\mathbf{r},t') = (2\pi)^{-3/2} \int g(\mathbf{k})\, e^{-ak^2 t'/b}\, e^{-i\mathbf{k}\cdot\mathbf{r}}\, d^3k, \quad (16.50)$$

whence

$$g(\mathbf{k}) e^{-ak^2 t'/b} = (2\pi)^{-3/2} \int \phi(\mathbf{r}',t')\, e^{i\mathbf{k}\cdot\mathbf{r}'}\, d^3x'. \quad (16.51)$$

Substituting this $g(\mathbf{k})$ into Equation 16.49 and reversing the order of integrations, we have

$$\phi(\mathbf{r},t) = \int \mathcal{G}(\mathbf{r} - \mathbf{r}', t - t')\, \phi(\mathbf{r}',t')\, d^3x, \quad (16.52)$$

where

$$\begin{aligned} \mathcal{G}(\mathbf{R},T) &= \frac{1}{(2\pi)^3} \int e^{-ak^2 T/b - i\mathbf{k}\cdot\mathbf{R}}\, d^3k \\ &= [b/4\pi a(t - t')]^{3/2}\, e^{-b(\mathbf{r} - \mathbf{r}')^2/4a(t-t')}. \end{aligned} \quad (16.53)$$

The function $\mathcal{G}$ is, except for a factor b, the same as the Green's function G, but its use is different. G, when multiplied by the source function f and integrated over all space *and* time, gives the response of the system to f, whereas $\mathcal{G}$, sometimes called a *propagator,* is multiplied by ϕ at *one* time and integrated over all space only, to give ϕ at later times. In calculating G we found that it has to vanish for $t < t'$, that is, a cause cannot produce an earlier effect. No such restriction has been discovered for $\mathcal{G}$ in the foregoing argument, but if a/b has a positive real part the integral in Equation 16.53 diverges for negative T. In the diffusion and heat-flow equations, a and b are both real and positive, so the propagator $\mathcal{G}$ can be used only for $t > t'$. This restriction is a consequence of the fact that diffusion and heat flow are inherently dissipative processes, so their equations of motion are not time-reversible in the sense of Chapter 8. Note that taking a negative value of $t - t'$ renders $\mathcal{G}$ complex, and clearly unacceptable. In particular, there is no $\mathcal{G}$ for the homogeneous form of Equation

16.37 that undoes the effect of the $\mathscr{G}$ in Equation 16.53. In the Schrödinger equation, b is imaginary and the sign of $t - t'$ does not affect the convergence of the integral. The Schrödinger equation in the form Equation 16.36 is time-reversible, provided one replaces ψ by ψ^*.

Our one-dimensional equations for periodic and continuous waves contain second time derivatives, and keep these derivatives when extended to three dimensions. In the analysis of vibrations of crystal lattices (and in some analyses of quantum field theories) one studies problems like that of the chain of masses, but with each mass connected to neighbors in all directions—sometimes also to second-nearest neighbors. Such problems are very difficult and only the simplest have been precisely solved. The three-dimensional extension of the continuum wave equation is more commonly encountered and is dealt with by a great number of mathematical techniques such as separation of variables, Green's functions, and "Fourier" integrals. The word Fourier is in quotation marks because the functions being (continuously) summed in such integrals are not necessarily sinusoidal; functions of time are likely to be, but functions of position are determined by the geometry of the coordinate systems being used. Physical situations involving three-dimensional wave equations of second order include those in which sound waves and electromagnetic waves are propagated. The electromagnetic case is particularly complicated because the dependent variables in the wave equations are themselves vector functions instead of single-component functions such as string displacement in a single direction.

To study the three-dimensional extension of the continuous wave equation $\phi_{xx} = \phi_{tt}/v^2$, we include a source term also, getting the inhomogeneous *d'Alembert* wave equation:

$$\nabla^2\phi - \phi_{tt}/v^2 = -4\pi f, \tag{16.54}$$

where ϕ and f are assumed to depend on x,y,z,t. To get a Green's function for this equation we again use the expressions in Equations 16.38 and 16.39. These, with Equation 16.54 imply that

$$\beta = \frac{-4\pi v^2\alpha}{(\omega - vk)(\omega + vk)}. \tag{16.55}$$

We substitute this expression in Equation 16.38, express α by Equation 16.39, and reverse the order of integration:

$$\phi(\mathbf{r},t) = \int G(\mathbf{r}-\mathbf{r}',t-t')\, f(\mathbf{r}',t')d^3x'dt', \tag{16.56}$$

where

$$G(\mathbf{R},T) = \frac{-v^2}{4\pi^3}\int \frac{e^{i(\omega T-\mathbf{k}\cdot\mathbf{R})}}{(\omega - vk)(\omega + vk)}\, d^3k d\omega. \tag{16.57}$$

In doing the ω integration we encounter an ambiguity: the poles are on the real axis and can be treated as being in either of the semicircles, or, via Cauchy principal values, each pole can be split between the semicircles. We keep our options open by including in the upper semicircle the residue at ck, times a coefficient **a** between 0 and 1, plus that at $-$ck times **b**, also between 0 and 1. Thus, when $T > 0$, using the upper semicircle, we get

$$G(\mathbf{R},T) = \int \frac{-ivd^3k}{4\pi^2k}\,[ae^{i(vkT-\mathbf{k}\cdot\mathbf{R})} - be^{i(-vkT-\mathbf{k}\cdot\mathbf{R})}]; \tag{16.58}$$

when $T < 0$, using the lower semicircle, we get

$$G(\mathbf{R},T) = \int \frac{ivd^3k}{4\pi^2k}[(1-a)e^{i(vkT-\mathbf{k}\cdot\mathbf{R})} - (1-b)e^{i(-vkT-\mathbf{k}\cdot\mathbf{R})}]. \tag{16.59}$$

We now do the **k** integration by using spherical coordinates in **k**-space, k,θ,ψ, with the volume element $k^2 \sin\theta\, dkd\theta d\psi$. We choose the polar axis to lie in the direction of the vector $\mathbf{R} = \mathbf{r} - \mathbf{r}'$, so $\mathbf{k}\cdot\mathbf{R} = kR\cos\theta$. The integrands are independent of ψ, so that part of the integration can be done immediately, yielding a factor 2π. Then

$$\int_0^{2\pi} d\psi \int_0^{\pi} \sin\theta e^{-i\mathbf{k}\cdot\mathbf{R}}\, d\theta = \frac{2\pi i}{kR}\,(e^{-ikR}-e^{ikR}). \tag{16.60}$$

Then, if $T > 0$,

$$G(\mathbf{R},T) = \frac{v}{2\pi R}\int_0^\infty dk[a(e^{ik(vT-R)} - e^{ik(vT+R)})$$

$$- b(e^{-ik(vT+R)} - e^{-ik(vT-R)})] \qquad (16.61)$$

and, if $T < 0$,

$$G(\mathbf{R},T) = \frac{-v}{2\pi\ R}\int_0^\infty dk[(1-a)(e^{ik(vT-R)} - e^{ik(vT+R)})$$

$$- (1 - b)(e^{-ik(vT+R)} - e^{-ik(vT-R)})] \qquad (16.62)$$

If $a = b$, each of these integrals can be rewritten as an integral from $-\infty$ to $+\infty$, and then evaluated in accordance with Equation 14.44; for $T>0$,

$$G(\mathbf{R},T) = \frac{va}{R}[\delta(vT - R) - \delta(vT + R)] = \frac{va}{R}\delta)(vT - R); \qquad (16.63)$$

for $T<0$,

$$G(\mathbf{R},T) = \frac{v(1-a)}{R}[\delta(vT + R) - \delta(vT - R)] = \frac{v(1-a)}{R}$$
$$\delta(vT + R). \qquad (16.64)$$

In each case we have dropped the second delta function in the square brackets, because these functions are zero throughout the indicated ranges of T. Now, also, we can write $v\ \delta(vT \pm \mathrm{R}) = \delta(T \pm \mathrm{R}/v)$, as one can prove by using the delta functions in an integral. Then we interpret these results by expressing the function $\phi(\mathbf{r},t)$:

$$\phi(\mathbf{r},t) = \int d^3x'$$
$$\times \frac{af[\mathbf{r}',t - (|\mathbf{r} - \mathbf{r}'|)/v] + (1 - a)f[\mathbf{r}',t + (|\mathbf{r} - \mathbf{r}'|)/v]}{|\mathbf{r} - \mathbf{r}'|} \qquad (16.65)$$

Here we have used the delta functions to eliminate the t' integration; their effect is to annul all contributions to ϕ from the source f *except* at the time at which a wave of speed v would have had to start from $\mathbf{r}'$ to reach $\mathbf{r}$ at time t, *and* the times at which a wave starting from $\mathbf{r}$ at time t would have arrived at $\mathbf{r}'$. These two parts of the integral, with coefficients a and $1 - a$ respectively, are called *retarded* and *advanced* solutions of Equations 16.54.

In all (or nearly all) applications, one has to choose $a = 1$, so only the retarded solution is present; effects follow causes, so if f causes ϕ, its contribution to ϕ must originate at time $t - R/v$, before its effect is felt. There have been some treatments of electrodynamics in which both retarded and advanced potentials (which satisfy Equation 16.54) were used, but they present problems of interpretation.

Aside from describing electromagnetic fields and potentials, Equation 16.54 is applicable to mechanical waves such as sound. It implies nondispersive propagation, as one can see either by applying it to an infinite plane wave (in which case it reduces to the string equation), or by noting that the waves spreading from various points in the source all travel from $\mathbf{r}'$ to $\mathbf{r}$ with the same speed v, regardless of the nature of the time dependence of f.

It is interesting to seek a propagator like $\mathcal{G}$ for Equation 16.54 with $f = 0$. Because the equation is of second order in t, its solution is not fully determined by ϕ as a given function of $\mathbf{r}$ at a fixed time; as in the case of the string, one needs also to know ϕ_t as a function of $\mathbf{r}$ at that time. Thus an equation that plays the role of Equation 16.52 will necessarily be more complicated than Equation 16.52. Substituting from Equation 16.38 into Equation 16.54 with $f = 0$, we find that ω is ambiguous as a function of k:

$$\omega = \pm vk. \tag{16.66}$$

Thus, although our quadruple integral reduces to a triple integral as in Equation 16.49, there are two functions $g(\mathbf{k})$ to determine:

$$\phi = (2\pi)^{-3/2} \int d^3k \left[g_+(\mathbf{k})e^{i(vkt-\mathbf{k}\cdot\mathbf{r})} + g_-(\mathbf{k})e^{i(-vkt-\mathbf{k}\cdot\mathbf{r})}\right]. \tag{16.67}$$

We cannot determine these two functions of $\mathbf{k}$ from one function $\phi(\mathbf{r})$ at one time; we need another function of $\mathbf{r}$ also, such as the time derivative of ϕ, $\phi_t(\mathbf{r})$, at that same time:

$$\phi_t = iv(2\pi)^{-3/2} \int kd^3k\,[g_+(\mathbf{k})e^{i(vkt-\mathbf{k}\cdot\mathbf{r})} - g_-(\mathbf{k})e^{-i(vkt+\mathbf{k}\cdot\mathbf{r})}] \tag{16.68}$$

$$g_+e^{ivkt'} + g_-e^{-ivkt'} = (2\pi)^{-3/2} \int \phi(\mathbf{r}',t')e^{i\mathbf{k}\cdot\mathbf{r}'}\,d^3x' \tag{16.69}$$

$$g_+e^{ivkt'} - g_-e^{-ikvt'} = \frac{-i(2\pi)^{-3/2}}{vk} \int \phi_{t'}\,(\mathbf{r}',t')e^{i\mathbf{k}\cdot\mathbf{r}'}\,d^3x',$$

Thus

$$g_+ = \frac{1}{2(2\pi)^{3/2}} \int [\phi(\mathbf{r}',t') - \frac{i}{vk}\,\phi_{t'}\,(\mathbf{r}',t')]e^{i(\mathbf{k}\cdot\mathbf{r}'-vk\,t')}\,d^3x' \tag{16.70}$$

$$g_- = \frac{1}{2(2\pi)^{3/2}} \int [\phi(\mathbf{r}',t') + \frac{i}{vk}\,\phi_{t'}\,(\mathbf{r}',t')]e^{i(\mathbf{k}\cdot\mathbf{r}'+vk\,t')}\,d^3x'$$

We substitute these expressions in Equation 16.67 and, as usual, reverse the order of integration:

$$\phi(\mathbf{r},t) = \frac{1}{16\pi^3} \int d^3x'\,\{(\phi(\mathbf{r}',t') \int d^3ke^{-i\mathbf{k}\cdot\mathbf{R}}\,(e^{ivkT} + e^{-ivkT})$$

$$- \frac{i}{v}\,\phi_{t'}\,(\mathbf{r}',t') \int \frac{d^3k}{k}\,e^{-i\mathbf{k}\cdot\mathbf{R}}\,(e^{ivkT} - e^{-ivkT})\} \tag{16.71}$$

$$= \int d^3x'\,[\mathcal{G}(\mathbf{R},T)\,\phi_{t'}\,(\mathbf{r}',t') - \mathcal{G}_{t'}\,(\mathbf{R},T)\,\phi(\mathbf{r}',t')]$$

where

$$\mathcal{G}(\mathbf{R},T) = \frac{i}{16\pi^3 v} \int \frac{d^3k}{k}\,e^{-i\mathbf{k}\cdot\mathbf{R}}\,(e^{ivkT} - e^{-ivkT}). \tag{16.72}$$

Here, as before $\mathbf{R} = \mathbf{r}-\mathbf{r}'$ and $T = t - t'$. We evaluate the integral as we did those in Equations 16.58 and 16.59:

$$\mathcal{G}(\mathbf{R},T)$$

$$= \frac{1}{8\pi^2 vR}\int_0^{\infty} dk[e^{ik(vT-R)} - e^{ik(vT+R)} - e^{ik(-vT-R)} + e^{ik(-vT+R)}]$$

$$= \frac{1}{4\pi vR}[\delta(vT-R) - \delta(vT+R)]. \tag{16.73}$$

This function is the coefficient of the time derivative of ϕ in the integral in Equation 16.71, and its time derivative is the coefficient of ϕ in that integral

$$\mathcal{G}_{t0}(\mathbf{R},T) = \frac{1}{4\pi R}[\delta'(vT + R) - \delta'(vT - R)]. \tag{16.74}$$

The integral in Equation 16.71, although expressed as an integral over all space, actually reduces to a surface integral on account of the delta functions. A choice of a sign of T (for example, a decision to predict ϕ at a given point from its values at an earlier time or over a range of earlier times) determines which delta function in $\mathcal{G}$ and in $\mathcal{G}_{t0}$ will contribute: in the case of prediction, $T > 0$ and the second term in each function will be the effective one and the first terms can be dropped. Equation 16.54 being time-reversible without f, however, retrodiction is feasible, also. In either case a value chosen for T in a given direction from the point $\mathbf{r}$ determines a value of R in that direction, and the set of all such choices determines a closed surface surrounding the point $\mathbf{r}$. The integral over all space reduces to an integral over that surface. It is convenient to integrate by parts the terms arising from Equation 16.74, so that one ultimately finds oneself integrating a radial derivative and a time derivative of ϕ over a closed surface at a set of times proportional to the various distances of the surface from the point $\mathbf{r}$. If all the times are chosen equal, the surface is a sphere with its center at $\mathbf{r}$.

If one applies either Equation 16.37 or Equation 16.54 to functions f and ϕ that are assumed to be independent of time, one has Poisson's equation

$$\nabla^2\phi = -4\pi f, \tag{16.75}$$

which reduces to Laplace's equation when $f = 0$:

$$\nabla^2\phi = 0. \tag{16.76}$$

Equation 16.65 shows that the Green's function for Poisson's equation is simply $1/|\mathbf{r} - \mathbf{r}'|$, which yields the familiar expression for the electrostatic potential produced by a stationary charge distribution. The procedures used above, however, do not work on Laplace's equation, for substituting a Fourier integral for ϕ in that equation gives $\phi = 0$ or $k = 0$. This means that no nontrivial solution of Laplace's equation can be expressed as a Fourier integral. However, one can by other arguments express ϕ at a point in terms of an integral, over a surface surrounding the point, of a function involving ϕ and its derivative normal to the surface.

NOTES

1. The behavior of gases in adiabatic change is discussed in most books of thermodynamics, for example, Zemansky M.W. and Dittman, R.H. *Heat and Thermodynamics,* 6th ed., New York: McGraw Hill, 1981.
2. A good general reference on the equations discussed in this chapter is Morse, P.M. and Feshbach, H. *Methods of Theoretical Physics.* New York, McGraw Hill, 1953.

PROBLEMS

16.1 Work out expressions for energy-flow density and kinetic-energy density for a plane sound wave of small amplitude. Then derive the wave equation from Hamilton's principle (Equation 14.15), and prove the equation of continuity (Equation 14.25).

16.2 Calculate a Green's function for Equation 16.37, modified by the addition of a term $c\phi$ on the left.

16.3 If, in Equation 16.37, $f(\mathbf{r},t) = 0$ and $\phi(\mathbf{r},0) = Ae^{-Br^2}$, determine $\phi(\mathbf{r},t)$.

16.4 For physical reasons, absolute temperature and concentration of a solute must be nonnegative and real. Prove that, if a solution of

Equation 16.37 has these properties at one time, it has them at all times.

16.5 Assuming $t > t' > t''$, prove that the $\mathcal{G}$ in Equation 16.53 has the property that

$$\int \mathcal{G}(\mathbf{r} - \mathbf{r}', t-t')\, \mathcal{G}(\mathbf{r}' - \mathbf{r}'', t' - t'')\, d^3x' = \mathcal{G}(\mathbf{r} - \mathbf{r}'', t - t'').$$

Why is this an essential property for $\mathcal{G}$ to have?

17

Nonlinear Waves

Having now completed our discussion of continuous and periodic linear systems, of linear wave equations in particular, we shall now, as in Chapter 8, have a glimpse of nonlinear waves. The study of such waves is at present a large and active field of study in applied mathematics. The activity is motivated largely by the fact that the most fundamental equations of physics are nonlinear—the wave equations of general relativity and of gauge theories of particles—and by the fact that some solutions of certain nonlinear wave equations behave rather like particles even before they are subjected to quantum postulates. Partly because these particle properties appear only in the one-dimensional versions of some of these equations and partly because such versions are much simpler to analyze than more realistic versions, we shall confine our discussion to equations containing two independent variables, x and t.

We begin with a study of the ladder network of which one section is depicted in Figure 17.1. The dependence of the circuit elements on D, the length of one section of the network, is so chosen that L, C_1, and C_2 are the parameters associated with a unit

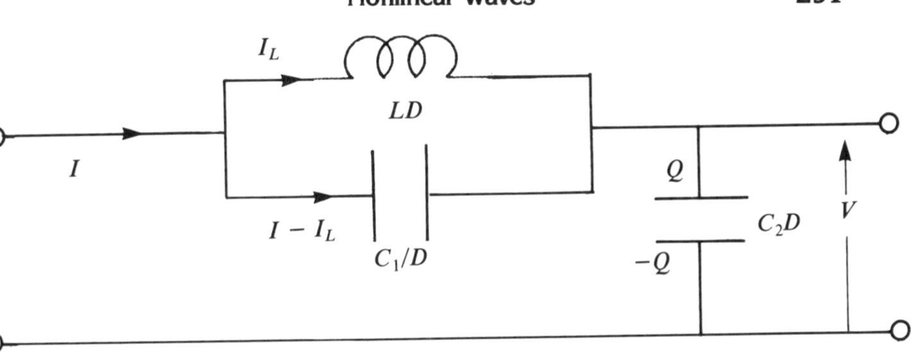

Figure 17.1. One section of a ladder network that can behave according to the Korteweg-de-Vries equation.

length. On the assumption that all these elements are linear, we can analyze the network by the methods of Chapter 13. This analysis tells us that the network is a low-pass filter whose cut-off frequency ω_0 is given by

$$\frac{1}{\omega^2} \geqslant \frac{1}{\omega_0^2} = \frac{1}{\omega_1^2} + \frac{1}{4\omega_2^2} \equiv LC_1 + \tfrac{1}{4}LC_2D^2. \qquad (17.1)$$

In the low-frequency limit the phase velocity of waves on the network is

$$v_\phi \underset{\omega \to 0}{\longrightarrow} (LC_2)^{-1/2} = D\omega_2. \qquad (17.2)$$

The phase velocity depends on frequency throughout the pass band, so any pulse will change as it travels; propagation is dispersive. Wavelength varies from ∞ at $\omega = 0$ to $2D$ at ω_0.

We introduce a nonlinear relationship between Q and V,

$$Q = C_2V - C_2'V^2, \qquad (17.3)$$

and assume wavelengths long enough to permit replacing difference by differential equations for voltage, charge, and currents. Initially we leave open the possibility of a relationship different from Equation 17.3:

$$DL\,\frac{\partial I_L}{\partial t} = -D\,\frac{\partial V}{\partial x};$$

$$D\,\frac{I - I_L}{C_1} = -D\,\frac{\partial^2 V}{\partial x dt}; \tag{17.4}$$

$$\frac{\partial Q}{\partial t} = -D\,\frac{\partial I}{\partial x}.$$

The first two of these express the usual linear voltage-current relationships for L' and C, and the third expresses charge conservation. We have written $D\partial/\partial x$ for a difference of something between adjacent sections of the network. Now we eliminate all currents from Equation 17.4, getting a relationship involving only V and Q:

$$V_{xxtt} + \omega_1{}^2 V_{xx} = \frac{1}{DC_1}\,Q_{tt}, \tag{17.5}$$

where we have reverted to the subscript notation for partial derivatives. It is convenient to use a new time variable:

$$T = t/\sqrt{LC_2}. \tag{17.6}$$

Then insertion of Equation 17.3 into Equation 17.5 yields

$$C_1 V_{xxTT}/C_2 + V_{xx} - V_{TT} + C_2'(V^2)_{TT}/C_2 = 0. \tag{17.7}$$

Now we transform to a frame of reference moving toward $+x$ with the velocity given in Equation 17.2, and introduce a scaling parameter ε:

$$\xi = \varepsilon^{1/2}(x - T) = \varepsilon^{1/2}\left(x - \frac{t}{\sqrt{LC_2}}\right);$$

$$\tau = \varepsilon^{3/2} T = \varepsilon^{3/2}\,\frac{t}{\sqrt{LC_2}}. \tag{17.8}$$

We use the same parameter ε in expressing V:

$$V(\xi,\tau) = \varepsilon\mu_0(\xi,\tau) + \varepsilon^2\mu_2(\xi,\tau) + \ldots, \tag{17.9}$$

The purpose of these rewritings is to render an intractable equation manageable, and in particular to facilitate the search for waves of long wavelength. The new frame of reference is the rest frame of long waves on the linear network, and the new time scale gives these waves unit velocity with respect to the network. The parameter ε permits the solution V to be found by successive approximations (in a manner reminiscent of Picard's method as it was used in Chapter 8) in a form that is particularly adapted to long-wavelength solutions. A small value of ε produces rapid convergence of the series for V and, according to Equation 17.8, reduces a large interval of $x-T$ to a smaller interval of ξ. Also

$$\frac{d\xi}{d\tau} = \frac{1}{\varepsilon}\left(\sqrt{LC_2}\,\frac{dx}{d\tau} - 1\right) = \frac{1}{\varepsilon}\left(\frac{dx}{dT} - 1\right)$$

or (17.10)

$$\frac{dx}{dt} = \frac{1}{\sqrt{LC_2}}\left[\varepsilon\,\frac{d\xi}{d\tau} + 1\right].$$

Thus a small value of $dx/dT - 1$, the difference of velocity from the limiting long-wavelength velocity on the linear network, becomes a larger value of $d\xi/d\tau$ when ε is small. This argument suggests *assuming* that ε is small enough to produce rapid convergence of Equation 17.9, and checking the assumption by looking for solutions with values of $d\xi/d\tau$ and wavelength of order of magnitude one.

We change independent variables in Equation 17.7 and substitute the power series Equation 17.9 for V. Then we require the resulting equation to be valid for every value of ε for which Equation 17.9 converges. This requirement implies that the coefficient of *each* power of ε has to vanish. The lowest power that is present is the cube; the coefficient of ε^3 is

$$\frac{C_1}{C_2}\,\mu_{1_{\xi\xi\xi}} + 2\mu_{1_{\xi\tau}} + \frac{C_2'}{C_2}\,(\mu_1^2)_{\xi\xi} = 0. \tag{17.11}$$

Noting that each term contains a ξ-derivative, we reduce the order of the equation by defining $u=(u_1)_\xi$; then

$$u_\tau + \frac{C_2'}{C_2} uu_\xi + \frac{C_1}{2C_2} u_{\xi\xi\xi} = 0. \tag{17.12}$$

An equation of this form is known as the Korteweg–deVries (KdV) equation. By the substitutions

$$\xi = \left(\frac{2C_2}{C_1}\right)^{1/5} y; \; u = - \frac{6x2^{1/3}C_2^{4/3}}{C_2'C_1^{1/3}} w,$$

we can put the equation in one of its more usual forms:

$$w_\tau - 6ww_y + w_{yyy} = 0. \tag{17.13}$$

The KdV equation was first derived by Korteweg and deVries in 1895 as a corrected version of the linear equation that describes waves on water in the case of amplitude $\ll$ depth and wavelength $\gg$ depth (the case studied in the wave's rest frame in the preceding chapter); Korteweg and deVries kept the condition on wavelength, but relaxed the requirement of small amplitude. The derivation is lengthy and will not be given here. We mention only that, like the network equation, it is expressed in the rest frame of long-wavelength linear waves, that it has higher order than second because it initially has two unknowns—elevation of the liquid surface and velocity of the liquid, and elimination of one of these requires additional differentiation of the other—and that the nonlinear term comes from the convective derivative of velocity (see Equation 16.19). Like the KdV equation for the network, it is a first correction to the linear approximation—sometimes called "weakly nonlinear." It has been shown that without the nonlinear term the equation would be linear and dispersive, whereas without the third derivative it describes waves that break like those on a beach. We shall see that the two terms together lead to the existence of pulses that do not change while traveling. The KdV equation has been derived also for other physical systems, e.g., waves in plasma and in the atmosphere.

As stated, the equation has some solutions representing waves

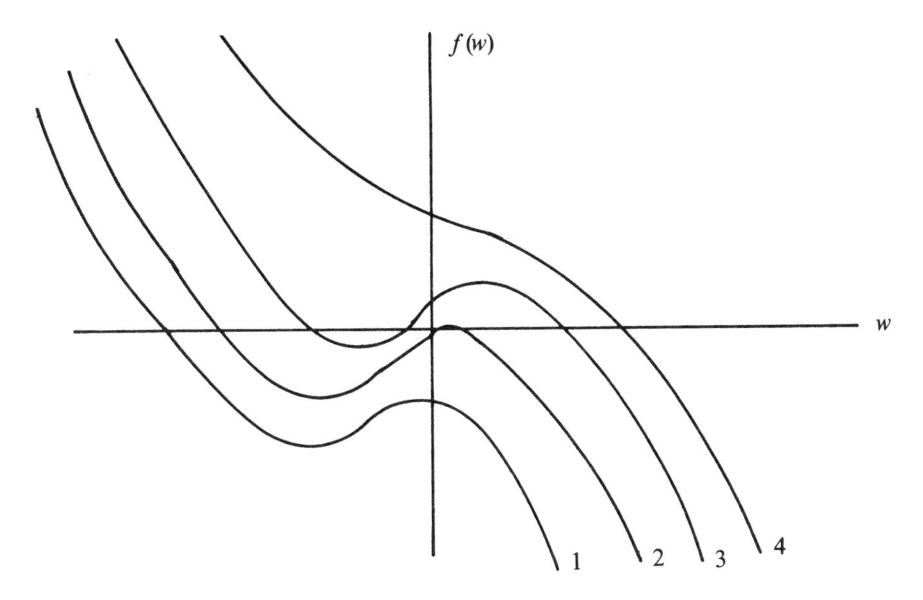

Figure 17.2. Typical graphs of f(w) vs. w, showing the effect of various combinations of parameters.

propagating nondispersively. To derive these solutions, we define $X = y - c\tau$, and try to find a solution that is a function of X alone (with c to be determined). Substituting $w(X)$ into Equation 17.13, we get

$$w''' = cw' + 3(w^2)', \tag{17.14}$$

where the prime means d/dX. Integrating once, we get

$$w'' = cw + 3w^2 + K, \tag{17.15}$$

where K is a constant of integration. This is of the form of a mechanical equation of motion, and can be integrated by being multiplied by w':

$$w'w'' = \tfrac{1}{2}(w'^2)' = (cw + 3w^2 + K)w' = (d/dX)\int (cw + 3w^2 + K)dw. \tag{17.16}$$

Thus

$$\tfrac{1}{2}w'^2 = \tfrac{1}{2}cw^2 + w^3 + Kw + K', \tag{17.17}$$

where K' is another constant of integration. We rewrite this as

$$\begin{aligned} \tfrac{1}{2}w'^2 + f(w) &= 0; \\ f(w) &= -w^3 - \tfrac{1}{2}cw^2 - Kw - K'. \end{aligned} \tag{17.18}$$

This relationship can be understood qualitatively by analogy with the classical mechanics of one-dimensional conservative systems (compare the first few equations of Chapter 8), $\tfrac{1}{2}w'^2$ is analogous to kinetic energy and $f(w)$ to potential energy, and the analogue of total energy equals zero. Typical graphs of $f(w)$ are shown in Figure 17.2. The range of values assumed by w is the interval between any two adjacent zeros of f in which interval f is negative. If negative values of f extend from a zero all the way to ∞, any value of w in that range is part of an unbounded function of X and is therefore physically inadmissible. Thus functions $f(w)$ represented by curves like 1 and 4 in Figure 17.2 are impossible. In the case of a curve like 3, it can be seen that w cannot have a value to the right of the most positive zero, for it would then be unbounded, but it can have values between the other two zeros. In this case w is a bounded periodic function of X whose range is the interval between the two zeros. Curve 2 has a double zero; if w lies to the left of this zero it is bounded, but nonperiodic.

We make these statements more quantitative by writing the solution of Equation 17.18 in the implicit form $X(w)$:

$$X = \int \frac{dw}{\sqrt{2w^3 + cw^2 + 2Kw + 2K'}}. \tag{17.19}$$

The integral is an elliptic integral of the first kind. If we label the zeros of $f(w)$ as w_1, w_2, w_3, counting from the right, and let X_3 be the value of X where $w = w_3$, we can make the equation more explicit:

$$X - X_3 = \int_{w_3}^{w} \frac{dw}{\sqrt{2w^3 + cw^2 + 2Kw + 2K'}}. \qquad (17.20)$$

X being an elliptic integral of w, w is an elliptic function of X. We shall not elaborate here on the properties of elliptic functions, but shall merely state that in general the solution above is expressible in terms of the (periodic) Jacobian elliptic function *cn*, and is called the "cnoidal" solution. Using conventional notation, we write

$$w(X) = w_2 - (w_2 - w_3)\, cn^2[\sqrt{\tfrac{1}{2}(w_2 - w_3)}\, (X - X_3)m]. \qquad (17.21)$$

The so-called "parameter" m is given by

$$m = (w_2 - w_3)/(w_1 - w_3). \qquad (17.22)$$

When $m < 1$, the *cn* function is rather like the cosine, and, as stated before, w is a periodic function of X, the cnoidal solution. The amplitude of this wave is $\frac{1}{2}(w_2 - w_3)$ and, as is generally true in a cubic polynomial, the velocity is

$$c = -2(w_1 + w_2 + w_3). \qquad (17.23)$$

One should remember that this is the amount by which the velocity exceeds that of long waves on the linear network, expressed in units of y, τ. The period of the wave, in X, is twice the integral in Equation 17.20 integrated from w_3 to w_2; this is the wavelength on the scale of y, or the frequency in τ units, divided by c. Clearly there are cnoidal waves with all velocities, wavelengths, frequencies, and amplitudes in the respective continuous ranges.

It is clear from Equation 17.13 that a complete solution is determined by the function $w(y)$ at a given value of τ—though this solution cannot be generated for a nonlinear equation by a Green's function or propagator. If $w(y)$ is a function of the form 17.21 at a given τ, the time-dependent solution will be of that form as a function of X, with the parameters c, K, K', or, equivalently, w_1, w_2, w_3, determined by the initial conditions.

Curve 2 depicts a limiting case in which $K = K' = 0$, or $w_1 = w_2$. Then Equation 17.20 becomes

$$X - X_3 = \int_{-c/2}^{w} \frac{dw}{w\sqrt{2w + c}} = -\frac{2}{\sqrt{c}} \tanh^{-1} \frac{\sqrt{2w+c}}{\sqrt{c}},$$

whence (17.24)

$$w = -\tfrac{1}{2}c \operatorname{sech}^2 \frac{\sqrt{c}}{2}(X - X_3).$$

This is a nonperiodic wave of the type called a "solitary" wave; like the periodic cnoidal wave, it propagates without changing shape. It is a negative pulse that goes to zero exponentially at $\pm\infty$. Its speed c (in y,τ units) relative to the motion of long wavelengths on the linear network, is necessarily positive, and the amplitude and width of the pulse are both related to c. Large values of c imply large (negative) amplitude and small width. Because the variable w is proportional to $-u$ in Equation 17.13, the solitary wave is actually a positive voltage pulse. If we had not changed this sign in the transformation, the KdV equation would have had a positive nonlinear term and solutions that would be positive solitary waves. These waves also are determined by $w(y)$ at some one time; if w is the kind of function in Equation 17.24 initially, it will remain such a function of X. If, at one time, $w(y)$ is neither the cnoidal nor the solitary-wave function, it will, according to the KdV equation, propagate with changing shape. Incidentally, it may be worth mentioning that a sinusoidal $w(y)$ that is small enough to permit the nonlinear term to be ignored will also propagate with fixed shape, for then the KdV equation is linear;

$$w_\tau + w_{yyy} = 0. \qquad (17.25)$$

This equation implies dispersive propagation with negative phase velocity and negative group velocity. These backward (relative to the moving frame of reference) small ripples are then a third kind of wave that can travel with fixed shape.

The imprecision and incompleteness of our treatment is berayed by the fact that the KdV equation is not invariant under reversal of either τ or y. That is why, for example, the solitary waves can travel only faster in the $+y$ direction than the long linear waves; c cannot be negative. But the symmetry of the network (or the other physical systems to which the KdV equation has been applied) implies that any wave that can go toward $+y$ with a given speed can go also toward $-y$ with the same speed. These reversed waves are not solutions of the KdV equation. A clue to this inadequacy of the equation is found in the derivation that led to it, especially Equations 17.8 and 17.9. It was here that we assumed a velocity with respect to the moving frame that was small relative to the velocity of the frame, so that the resultant velocity had to be positive. We could equally well have derived a KdV equation for waves moving in the negative direction, but at the cost of losing our results for positive velocity. We do not have one manageable equation that encompasses both directions of propagation.

Despite such limitations of its application, the KdV equation has some remarkable properties which have stimulated much work on this equation and others that have similar properties. Here we can only sketch and summarize some of these relationships; the references at the end of the chapter contain bibliographies and fuller explanations.

The special properties of some partial differential equations are revealed by means of *Bäcklund transformations,* the first of which was discovered by Bäcklund in 1875 in connection with a problem in differential geometry. A Bäcklund transformation is a set of first-order PDEs which imply the more difficult PDE that one seeks to solve. For example, the Cauchy-Riemann conditions for analyticity (Chapter 5) are a Bäcklund transformation: $u_y = -v_x$, $u_x = v_y$ imply Laplace's equation for u and v by an argument that equates the second mixed derivatives taken in opposite order; $u_{xy} = u_{yx}$ and $v_{xy} = v_{yx}$. In this argument the $C-R$ equations express the Bäcklund transformation, and Laplace's equation plays the role of the "difficult" PDE that one seeks to solve. One does not normally need this kind of help in solving Laplace's equation, but it can be welcome in connection with some nonlinear PDEs.

The utility of the method is more apparent in connection with Liouville's equation:

$$u_{zz} - u_{tt} = 4e^u,$$

or (17.26)

$$u_{xy} = e^u,$$

where

$$x = z + t, y = z - t.$$

Tackling the equation in its xy form, we write the Bäcklund transformation

$$u_x + w_x = \sqrt{2}\, e^{(u-w)/2}; \tag{17.27}$$

$$u_y - w_y = \sqrt{2}\, e^{(u+w)/2}. \tag{17.28}$$

Now, differentiating these and substituting from them into the differentiated equation, we have:

$$\begin{aligned}(\text{Equation } 17.27)_y &\rightarrow (u+w)_{xy} = e^u \\ (\text{Equation } 17.27)_y - (17.28)_x &\rightarrow w_{xy} = 0\end{aligned} \tag{17.29}$$

Thus Equations 17.27 and 17.28 imply 17.26 and 17.29. This relationship is useful in solving Equation 17.26 because the general solution of Equation 17.29 is well known (see Chapter 14):

$$w = g(x) + f(y), \tag{17.30}$$

where f and g are arbitrary. This solution can be inserted into Equations 17.27 and 17.28, which can then be easily integrated to give the general solution of Equation 17.26 in terms of the arbitrary functions f and g.

The transformation that helps solve the KdV equation does not at first sight look like a Bäcklund transformation; it is known as *Miura's* transformation. We write the KdV equation with x,t as independent variables:

$$u_t - 6uu_x + u_{xxx} = 0. \tag{17.31}$$

Then Miura defined

$$w_x + w^2 = u; \tag{17.32}$$

$$w_t = 6w^2 w_x - w_{xxx} = (2wu - u_x)_x. \tag{17.33}$$

The second form of Equation 17.33 was derived from the first with the help of Equation 17.32. Substituting Equation 17.32 into Equation 17.31 leads to

$$u_t - 6uu_x + u_{xxx} = [2w + \partial/\partial x][w_t - 6w^2 w_x + w_{xxx}], \tag{17.34}$$

so if the last factor vanishes (i.e., if Equation 17.33 holds), u satisfies the KdV equation. The vanishing of that last factor, or Equation 17.33 in its first form, is known as the modified Korteweg-deVries (mKdV) equation, which differs from the KdV only by the square replacing the first power in the nonlinear term. The converse of this implication does not hold: Equation 17.31 + Equation 17.32 does not imply Equation 17.33.

Because the mKdV is as forbidding as the KdV itself, the relationship shown above is not directly useful in solving the KdV. But related arguments are useful. Equation 17.32 is a Riccati equation, which can be linearized by the substitution

$$w = \frac{\psi_x}{\psi}, \tag{17.35}$$

whence

$$\psi_{xx} - u\psi = 0. \tag{17.36}$$

This linear equation for ψ can be generalized by transformation of the KdV equation to a moving frame of reference:

$$\begin{aligned} x' &= x - at; \\ t' &= t. \end{aligned} \tag{17.37}$$

It is obvious that $u_{x'} = u_x$; $u_t = u_t' - au_{x'}$. Thus the KdV equation is valid in the new frame of reference if u is replaced by $u+a$, where a is the velocity of the new frame relative to the old. Thus, more generally,

$$\psi_{xx} - (u + a)\psi = 0. \tag{17.38}$$

This equation is related to the form of the Schrödinger equation shown as Equation 16.36, but with three differences: it is in one dimension instead of three; it assumes a time dependence of the form $e^{-i\omega t}$, which means physically that the particle has precise energy $\hbar\omega = -\hbar^2 a/2m$; it does not apply to a force-free particle, but to one with potential energy $\hbar^2 u(x)/2m$.

Equation 17.38 is the basis of the so-called *inverse scattering* method of solving the KdV equation. According to Equation 17.38 the function $u(x)$ affects the behavior of a wave function ψ. The time variable on which also u depends is not the time on which ψ depends; it enters Equation 17.38 as a parameter that determines what potential acts on ψ. If ψ is chosen to represent a wave incident from $x = -\infty$, the potential u will determine what fraction of the wave is reflected and what fraction transmitted. Because u depends not only on x but also on a parameter t, in general different values of this parameter will be associated with potentials that produce different amounts of reflection. It has been proved that, if $u(x,t)$ satisfies the KdV equation, then $u(x)$ depends on t in such a way that (among other things) the amount of reflection of ψ does not depend on t. Furthermore, if no reflection at all occurs, $u(x,t)$ is a solitary wave like that given in Equation 17.24, or a combination of such waves.

The inverse scattering method reverses what one usually does with the Schrödinger equation. Usually one assumes that the potential $u(x)$ is known, and seeks to find out how this potential affects the wave function ψ—in particular, how much of ψ is reflected. The inverse scattering method requires one to assume certain facts about the reflection, and from them to infer what the potential is—actually, a one-parameter family of potentials with parameter t. There are systematic approaches to this problem, but some of the relevant questions are still unanswered.

One very important consequence of these studies of the KdV

equation is that two solitary waves that interact—for example, a larger one that overtakes a smaller one—emerge from the interaction *unchanged,* except for a time delay or advance. Solitary waves that behave in this way are called *solitons.* Although their interaction is greatly affected by the nonlinear term in the KdV equation, they are like linear waves in being (almost) unaffected by passing through each other. Unlike linear waves, they do interact, but the interaction does not destroy the solitons as it might be expected to do in a nonlinear equation. This property of the equation has led physicists working on the theory of particles to study such equations as possible descriptions of elementary particles.

Another approach to the KdV equation uses a Bäcklund transformation to relate different solutions of the equation. This approach, which we do not present here, permits one to generate multisoliton solutions by successive transformations that start with $u=0$.

A nonlinear PDE that has been extensively studied is the *sine-Gordon* equation (SGE). The name comes from that of the Klein-Gordon equation, which is, in appropriately scaled units of distance and time,

$$\left(\nabla^2 - \frac{\partial^2}{\partial t^2}\right)\psi = 4\psi. \tag{17.39}$$

This equation has been much used as a relativistic wave equation; usually t is written as ct, and the number 4 is written as $m^2c^2/\hbar^2$, but these complications have been scaled away in Equation 17.39. For various reasons, it has been suggested that, instead of the linear equation 17.39, one should use

$$\left(\nabla^2 - \frac{\partial^2}{\partial t^2}\right)\psi = 4\,\sin\psi; \tag{17.40}$$

the temptation to call this equation the sine-Gordon equation was too strong to resist. Most of the work on the equation has been done on its one-dimensional form:

$$\psi_{xx} - \psi_{tt} = 4\,\sin\psi, \tag{17.41}$$

or

$$\psi_{uv} = \sin \psi, \tag{17.42}$$

where

$$u = x + t \tag{17.43}$$

and

$$w = x - t \tag{17.44}$$

are the independent variables defined in Chapter 14, and in this chapter in connection with Liouville's equation.

Figure 17.3 depicts a system that is described by the SGE (on the assumption that all wavelengths present are very long compared to the spacing of the rods supporting the weights). The upper zig-zag line is elastic with negligible mass, assumed to satisfy Hooke's law and thus to lead to the second derivative terms in Equation 17.41; the sin ψ term comes from the restoring forces acting on the pendulums, where ψ is the angle of the pendulum from the vertical. The lower straight line is a rigid rod around which the pendulum rods swing without friction. Both the horizontal rod and the elastic are assumed to be fixed at their ends. All the pendulums have similar light rigid rods, and all the bobs have the same mass.

If the support at one end is rotated through 360° and set down again, the system will have one twist in it, called a "kink." It is fairly clear that such a twist can travel along the chain of pendulums, but nothing can undo it except a second intervention from outside. This kink is the kind of solitary wave that the SGE predicts.

To see in a little more detail how the argument goes, we introduce the original Bäcklund transformation (the one that Bäcklund himself discovered):

$$\phi_u - \psi_u = 2k \sin \tfrac{1}{2}(\phi + \psi); \tag{17.44}$$

$$\phi_w + \psi_w = \frac{2}{k} \sin\tfrac{1}{2}(\phi - \psi). \tag{17.45}$$

If we form (Equation 17.44)$_w$ + (Equation 17.45)$_u$, and (Equation 17.44)$_w$ − (Equation 17.45)$_u$, and in each case use the equations themselves to eliminate first derivatives, we get

$$\psi_{uw} = \sin\psi;$$
$$\phi_{uw} = \sin\phi. \tag{17.46}$$

Thus the first-order Equations 17.44 and 17.45 imply the SGE for *both* functions. In practical terms this means that, having one solution to the SGE, we can substitute it into Equation 17.44 and 17.45 and calculate another solution.

We carry out this process, starting with the trivial solution $\psi = 0$. This implies, via Equations 17.44 and 17.45,

$$\phi_u = 2k\sin\tfrac{1}{2}\phi;$$
$$\phi_w = (2/k)\sin\tfrac{1}{2}\phi. \tag{17.47}$$

These equations are readily integrated:

$$2ku = 2\ln\tan\tfrac{1}{4}\phi + f(w);$$
$$2w/k = 2\ln\tan\tfrac{1}{4}\phi + g(u). \tag{17.48}$$

Here f and g are functions of integration, each playing the role in its own equation of a constant of integration and each arbitrary in its own equation. The two equations together, however, determine f and g up to a constant:

$$\tan\tfrac{1}{4}\phi = Ce^{ku+w/k}, \tag{17.49}$$

where C is an arbitrary real constant; thus

$$\phi(u,w) = 4\tan^{-1}(Ce^{ku+w/k}). \tag{17.50}$$

In terms of the original variables, x and t,

$$\phi = 4\tan^{-1}\left[Ce^{2(x-Ut)/\sqrt{1-U^2}}\right], \tag{17.51}$$

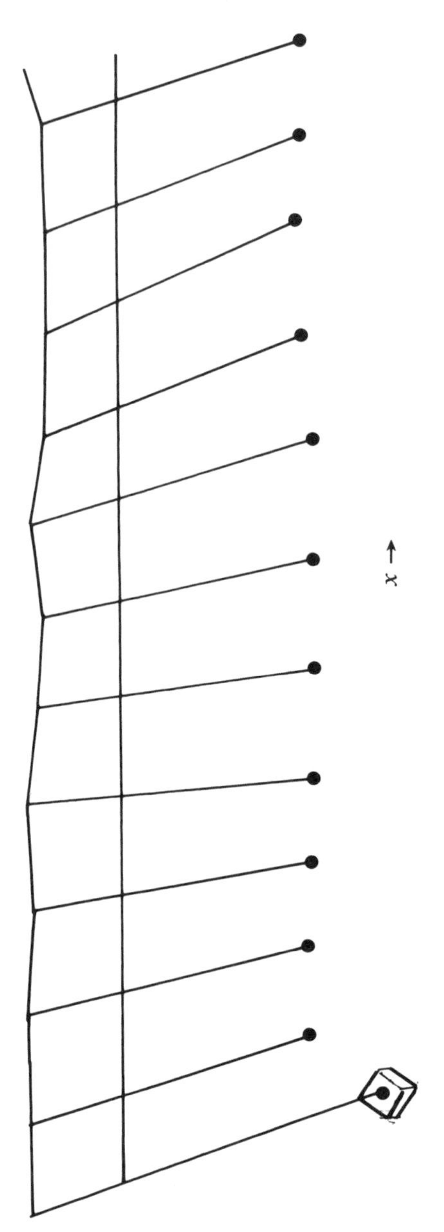

Figure 17.3. One end of a string of similar pendulums that are connected at their tops by elastic bands. This system is described by the sine-Gordon equation.

where

$$U = (k^2 - 1)/(k^2 + 1). \qquad (17.52)$$

This is a solitary wave that propagates with velocity U without changing shape. It must be interpreted as a wave on an infinite chain of pendulums, for no end conditions have been put in. Note that k can be any real number, so U ranges between -1 and $+1$. Notice, also, that at any given t, as x covers its infinite domain with fixed U, the exponential ranges from 0 to ∞, so the inverse tangent ranges from 0 to $\pi/2$, and ϕ ranges from 0 to 2π. This is one kink, which can travel on an infinite string at any speed from 0 to 1 in the present units, and in either direction.

If we now substitute this ϕ into the Bäcklund transformation and generate a new solution, it turns out to be a two-kink solution, etc. One interesting feature of these kinks which tantalizes particle theorists is that two kinks twisted in opposite directions can meet and annihilate each other, like antiparticles, so there is no kink in the final state.

In this chapter we have had a glimpse of a very large and active field of study. To have more than a glimpse would have been beyond the scope of the book. Readers are urged, however, to read the references mentioned here and to follow up this reading with an exploration of the extensive literature listed in the books' bibliographies.

NOTES

The following books have been useful in the writing of this chapter. Recommended references are:

1. Karpman, V.I. *Nonlinear Waves in Dispersive Media*. New York: Pergamon, 1975.
2. Lonngren, K. and Scott, A.C. eds. *Solitons in Action*. New York, Academic, 1978.
3. Eilenberger, G. *Solitons*. New York, Springer, 1981.

4. Dodd, R.K., J.C. Eilbeck, J.D. Gibbon, and H.C. Morris. *Solitons and Nonlinear Wave Equations.* New York; Academic Press, 1982.
5. Drazin, P.G. *Solitons.* Cambridge, Cambridge Press, 1983.

PROBLEMS

17.1 Complete the solution of Liouville's equation, Equation 17.26, and verify that it is a solution.

Answers to Selected Problems

Chapter 1

Problem 1.4.

$\omega_0 = 3.33 + 10^5 \text{ sec}^{-1}$,

$Q_0 = 7.75 + 10^{-4}$ coul.

Problem 1.11.

$2\pi/\omega_0$, π/ω_0.

Problem 1.12.

$\omega_0 = Q/\sqrt{mD^3} \text{ sec}^{-1}$, if Q, m, D are in cgs units.

Problem 1.13.

$\omega_0 = \sqrt{g/2L}$.

Chapter 2

Problem 2.7.

$$\frac{2\pi}{\omega'} \lesssim t = \cos^{-1} \frac{\omega'^2-\delta^2}{\omega'^2+\delta^2}$$

$$\frac{\pi}{\omega'} \lesssim t = \cos^{-1} \frac{\delta^2-\omega'^2}{\delta^2+\omega'^2} .$$

Problem 2.10.

$$\frac{1}{2}\left[\frac{\ln(1-f)}{2\pi}\right]^2$$

Chapter 3

Problem 3.4.

$$x(t) = \frac{A}{k}(1 + \delta t)e^{-\delta t}$$

Problem 3.7.

$x = 10 + 1.675 \cos 6.28t,$

$02x = 10 - 0.0255 \cos 62.8t,$ in cgs units.

Chapter 4

Problem 4.3.

$\Omega = 2k.$

$$\alpha_n = \frac{2A}{\pi(4n^2 - 1)} .$$

Problem 4.6b.

$$\frac{\sqrt{2}A}{a} e^{-\omega^2/4\alpha^2} \cos \omega t_0.$$

Problem 4.9.

$$f(t) = \sum_{n=\infty}^{\infty} \frac{A}{2\sqrt{\pi\alpha}} e^{-\pi^2 n^2/\alpha + 2\pi i n t}$$

Chapter 5

Problem 5.1.

a, c, k analytic everywhere in R.
b, i, l nonanalytic, respectively, where $f=0$, $g=0$, and $f=g$.
d, e, f, g, h, j nonanalytic everywhere.

Chapter 6

Problem 6.3.

$x(t) = 0$ if $t < 0$.

If $t > 0$, $x(t) = \dfrac{Ac}{m\omega'} \times$

$\{[(\delta - b)^2 + c^2 + \omega'^2]^2 - 4c^2\omega'^2\}^{-1}\{2\omega'(\delta - b)$
$(e^{-\delta t}\cos \omega' t - e^{-bt}\cos ct) + [(\delta - b)^2 + c^2 - \omega'^2]$
$(e^{-\delta t}\sin \omega' t - e^{-bt}\sin ct)\}$

Problem 6.4.

$$x(t) = 0,\ t < 0,\ x(t) = \frac{Ae^{-\delta t}}{2m\omega'^2}(\sin \omega' t - \omega' t \cos \omega' t),\ t \geqslant 0.$$

Chapter 7

Problem 7.5.

$h(\omega)$ can be written as N/D, where $N = \omega^2 Z_{12} Z_{22}$, and $D = \omega^2[Z_{12}(Z_1 + Z_2 + Z_{22}) + Z_1(Z_2 + Z_{22})]$. These are two polynomials in ω; D must have higher degree than N. Thus, for instance, either L_{12} or L_{22} must equal zero, and $L_{12}(L_1 + L_2 + L_{22}) + L_1(L_2+L_{22}) \neq 0$. If the latter condition fails, then it is necessary that $R_{12}L_{22} + R_{22}L_{12} = 0$, and the coefficient of ω^3 in D is nonzero . . . etc.

Chapter 8

Problem 8.2.

There are circular orbits where the polynomial $1/g(r)$ has zeros. These orbits are stable and thus act as limit cycles if f, hence $1/g$, has positive slope at its zeros. If all the zeros of $1/g$ are simple, the slope of $1/g$ alternates at its successive zeros; e.g., if the slope is positive at a, it is negative at b, positive at c, etc. But a double root (or any root of even multiplicity) yields the same sign of f on both sides of the root. Such a root corresponds to an unstable orbit. Thus the stable orbits (limit cycles) occur at the roots of $1/g$ of odd multiplicity at which the slope of $1/g$ is positive immediately above and below the root.

Chapter 10

Problem 10.2.

For each column matrix r, let $Ar = s$, that is, there is a unique column matrix s so produced. Now $XAr = Ir = r = Xs$, so X undoes what A does to any column matrix. Now, $r = Xs$, so $Ar = s = AXs = Is$; that is, if $XA = I$ it follows that $AX = I$, also. A commutes with its own inverse. Now assume that A has two inverses, X and Y, so $AX = I = AY$. Thus $A(X - Y) = 0$. Multiply from the left by X (Y would do as well): $XA(X - Y) = 0 = I(X - Y) = X - Y$, so $X = Y$; the inverse is unique.

Problem 10.7.

Let x be an eigenvector of P with eigenvalue p: $Px=px$. But $P=P^2$ so $P^2x = Ppx = pPx = p^2x = px$. Thus $p = p^2$, so $p = 0,1$. Since P isn't the unit matrix, there is at least one zero eigenvalue; P reduces the corresponding eigenvector to zero, an operation that has no inverse; also one can show that P has a zero determinant;

it is singular, having no inverse. It collapses all vectors in the space into their projections on a subspace of fewer dimensions. Incidentally, $P^3 = PP^2 = PP = P^2 = P$, etc; P equals each of its powers.

Chapter 11

Problems 11.2 and 11.8.

Use momentum conservation for initial conditions.

Problem 11.4.

Define x,y as coordinates measured along the two rods in one of the directions in which the rods diverge more slowly. Using cgs units, set

$$V = \frac{-Q^2}{\sqrt{D^2 + x^2 + y^2 - 2xy\cos\theta}} \simeq -\frac{Q^2}{D}\left[1 - \frac{x^2 - 2xy\cos\theta + y^2}{2D}\right],$$

$$T = \tfrac{1}{2}m(\dot{x}^2 + \dot{y}^2).$$

Proceed from there, using standard methods.

Chapter 12

Problem 12.1.

Suggestion: Use what you know about the chain fixed at its ends to infer the nature of the normal modes in the present system. *Then* work out the frequencies.

Problem 12.3.

Normal modes are the same, *mutatis mutandis.* Frequencies and phase velocities are all multiplied by $1/(1 - S_0/D)$.

Chapter 13

Problem 13.4.

$U^N = 1$ (the unit matrix). Solve this equation for the frequencies that satisfy it. This condition is equivalent to the one in U^{-1} obtained by retrograde analysis.

Chapter 14

Problem 14.1.

Gaussian pulse propagates solely toward $+x$.

Chapter 15

Problem 15.3.

Reflected wave =

$$\frac{-\alpha^2 I \cos(\omega t + kx) - 2\alpha I \sin(\omega t + kx)}{4 + \alpha^2}$$

$$\alpha \equiv \frac{km}{\mu}.$$

Transmitted wave =

$$\frac{-4I\cos(\omega t - \beta x) - 2\alpha I \sin(\omega t - kx)}{4 + \alpha^2}$$

Problem 15.5.

The combined impedance of two strings emerging from the vertex, as seen at the vertex, is the *sum* of the separate impedances (common v_y, sum of forces F_y). Thus, to avoid reflection of a wave incident along string #1 we match impedances: $\sqrt{\mu_1 \cdot \mathcal{T}_1} = \sqrt{\mu_2 \cdot \mathcal{T}_2} + \sqrt{\mu_3 \cdot \mathcal{T}_3}$. This condition negates the corresponding conditions for nonreflection of waves incident along other strings.

Chapter 16

Problem 16.4.

Equations 16.42 and 16.45 show that $\phi(r,t)$ is given by an integral with nonnegative integrand.

Index